OHSAS 18001

Designing and Implementing an Effective Health and Safety Management System

Joe Kausek

Government Institutes
An imprint of
The Scarecrow Press, Inc.
Lanham, Maryland • Toronto • Plymouth, UK
2007

Published in the United States of America
by Government Institutes, an imprint of The Scarecrow Press, Inc.
A wholly owned subsidary of
The Rowman & Littlefield Publishing Group, Inc.
4501 Forbes Boulevard, Suite 200
Lanham, Maryland 20706
http://www.govinstpress.com/

Estover Road, Plymouth PL6 7PY
United Kingdom

British Library Cataloguing in Publication Information Available

Library of Congress Cataloging-in-Publication Data

Kausek, Joe, 1957–
   OHSAS 18001 : designing and implementing an effective health and safety
management system / Joe Kausek.
      p. ; cm.
   Includes bibliographical references and index.
   ISBN-13: 978-0-86587-199-1 (pbk. : alk. paper)
   ISBN-10: 0-86587-199-X (pbk. : alk. paper)
   1. Industrial hygiene—Standards.   2. Industrial safety—Methodology.
3. Industrial management.   I. Title.
[DNLM: 1. Safety Management—standards.   2. Occupational Health
Services—organization & administration.   3. Safety Management—methods.
WA 485 K21o 2007]
RC967.K38   2007
616.9'803068—dc   222007001816

Contents

Preface

The current concern over health care costs has never been greater. Companies are eliminating health care programs or substantially increasing worker contributions, resulting in frustrated and dissatisfied employees. Retiree health care plans are being abandoned as corporations file for bankruptcy and struggle to survive. Automakers in Detroit report that the cost of employee health care is the single largest expense adding to the cost of a new automobile. The rising cost of health care is threatening many of the world's most well-known corporations with extinction.

At the same time, employees are working longer hours as organizations slash their workforce to stay lean. Employees have a wider variety of tasks that they must perform and less time to complete them. Workplace stress is higher than it has ever been, because employees are asked to do more with less.

What can an organization do to counter this rising threat? Quite simply, it can practice proactive measures to keep its workers healthy, safe, and fit. Insurance premiums and workers' compensation costs are directly tied to workplace injuries and claims. Reducing injuries and claims is one way to contain the rising costs of health care. Another is to improve the overall health and fitness of the organization's employees, both on and off the job. Such benefits extend well beyond the worker's employment and into their retirement as well, substantially reducing long-term health care costs associated with retirement plans.

Finally, a healthy, safe, and fit workforce is a more productive and motivated workforce. Never before has the claim that *"the employee is the organization's most valuable asset"* been more true. Equipment, machinery, and buildings no longer make up the bulk of the organization's strategic assets, rather it is the intangible knowledge, skills, and innovation of the company's employees that serve as the foundation of strategic core competencies. Keeping these vital assets on the job and productive is a strategic imperative.

An occupational health and safety assessment series (OHSAS) 18001 health and safety management system (HSMS) can help an organization improve its workplace health and safety and lower its health care costs. This book describes how to design and implement a HSMS based on the OHSAS 18001 specification. Throughout this book the focus is on the actions needed to generate widespread corporate support for and involvement in health and safety performance improvement. To support the goal of an efficient OHSAS 18001 implementation, dozens of tools, checklists, procedure templates, and spreadsheets that I have developed and used during several successful implementations of OHSAS 18001 in client organizations are provided. These tools not only speed the design and implementation of 18001, they also provide for efficient and effective ongoing maintenance of the system. The goal is to minimize the overhead needed to manage and maintain the system so that more time and resources are available for actions that actually improve employee health and safety.

By no means do the chapters in this book or the tools on the CD-ROM represent the only or in some cases even the best way to implement an OHSAS 18001 management system in any specific company. Any feedback or suggestions on this book or the tools provided would be appreciated. I can be contacted through my website at www.joekausekassociates.com.

Part 1

OCCUPATIONAL HEALTH AND SAFETY MANAGEMENT SYSTEM OVERVIEW

Part 1 examines the structure and intent of the OHSAS 18001 management system specification. The section begins with a look at what any management system should do and then briefly reviews the structure of OHSAS 18001. Of particular note in chapter 1 is the section that discusses management system evolution from conformance through effectiveness and toward improvement.

The OHSAS 18001:1999 specification is addressed in detail in chapter 2. The structure, interactions, and basic principles of occupational health and safety management as defined by that specification and the benefits that an organization can expect to gain from implementing a comprehensive health and safety management system are explored.

CHAPTER 1

Management Systems Overview

In general, any management system is composed of three primary components:

- The core processes that focus on the primary purpose or outputs of the system and the processes that produce them. In an occupational health and safety (OHS) system it is the identification of workplace hazards, assessment of the risks associated with these hazards, and then eliminating or minimizing these hazards or their risk through the proper implementation of operational controls and improvements. The processes used to determine, assign, and monitor occupational health and safety improvement objectives also can be classified as a core process in that it directly contributes to the core purpose of the HSMS.
- The key supporting processes that provide direct inputs into the core processes or measure the results of the outputs. These can include the processes used to maintain an awareness of legal and other requirements, ensure competency and awareness of employees, provide an adequate infrastructure, communicate important health and safety information, and monitor and evaluate performance.
- The management system supporting processes, such as document control, record control, and internal auditing.

For a management system to be effective, all of these components must be aligned and perform as they are intended. While most of the attention is given to the core processes, failure to adequately monitor or control any of the supporting processes will impact the effectiveness of the overall system. Now we will turn our attention to what a management system is supposed to do.

The Purpose of Any Management System

One can say that the primary purpose of any management system is to implement the chosen strategy of an organization by focusing resources on areas critical for organizational success. For an OHSAS 18001 management system, this means implementing the commitments stated in the organization's policy statements, including its commitment to provide a safe and healthy workplace for its employees. It might also include programs needed to effectively manage its health and safety costs.

To meet this purpose, all management systems establish requirements and guidelines, which when followed provide reasonable assurance that the outputs from the system will be as expected (i.e., a healthy and safe workplace, improved safety performance, and stable or decreasing health care costs, etc.). The purpose of the management system standards is to provide these general requirements. It should be noted that the OHSAS 18001 standard is nonprescriptive; that is, it details what should be done but not necessarily how an organization should do it. Organizations then develop and implement internal systems, processes, and procedures that provide more detail on how these general requirements are met within their unique operating environment.

The general trend in management systems is to reinforce the need to align processes into integrated systems of processes, which are all focused on providing the highest value to the customer. In this sense, the primary customers of the HSMS are the organization's employees, with secondary customers including owners or shareholders, customers, insurers, worker families, and government agencies. Effective occupational HSMSs provide significant benefits to each of these customer groups. Conversely, poorly constructed and instituted HSMSs can present significant risk (negative value) to these same customers. Those charged with developing, implementing, and maintaining these systems must understand and be able to articulate these benefits and risks to enlist the support of both management and employees in the common goal of creating an effective and efficient management system. These potential benefits and risks are examined next to prepare the reader for the consensus building needed to develop an effective management system.

The Case for OHSAS 18001

The OHSAS 18001 model for health and safety management can provide significant benefits to an organization. The case to implement OHSAS 18001 is driven by the need to better protect the organization's most valuable asset—its people.

The movement from the industrial age to the information age and now to the knowledge age has made organizations more dependent on their employees. Although the outsourcing of skilled labor has become a fixture of the global economy, this has not changed the fact that employees, no matter where they are located, are a more important part of the organization's success. As a result, keeping those employees healthy and safe has become a greater priority.

This is especially true in the United States, because health care costs have spiraled upward and now threaten the existence of major corporations. Health and safety management has moved beyond a singular focus on safety toward a more holistic focus on employee health, safety, and overall well-being. Studies have proven that a healthy worker is a more productive and cost-effective worker. To make meaningful improvements in worker health, an organization must move beyond regulatory compliance (risk reduction) toward more polices based on prevention (hazard elimination). Because OHSAS 18001 focuses on moving beyond regulatory compliance, it can provide significant benefits in this regard. Controlling both the short- and long-term costs of health care is now a strategic priority.

Some of the statistics relating to U.S. workers supporting an aggressive health and safety management program follow:

- Approximately six thousand workers die each year from workplace accidents.
- Approximately fifty thousand deaths occur each year from workplace exposures.
- There are approximately six million nonfatal injuries reported each year. The number of unreported injuries may be much higher.
- The estimated cost to U.S. business from workplace accidents is $125 billion.
- Fires and explosions kill more than two hundred workers each year and injure more than five thousand.
- An average of one worker is electrocuted every day in the United States.
- Powered industrial truck accidents kill more than one hundred workers each year; 25 percent of which are attributed to inadequate training for the employee. Over thirty-six thousand workers are injured each year.
- Almost six million workers are exposed to blood-borne pathogens each year.
- More than thirty-two million employees are exposed to hazardous chemicals. There are an estimated 650,000 chemicals in use in the workplace today. Chemical exposure may cause or contribute to many serious health effects, such as heart ailments, central nervous system damage, kidney and lung damage, sterility, cancer, burns, and rashes.

- Slip-and-fall accidents are responsible for 15 percent of all workplace fatalities. They are second only to accidents involving powered industrial trucks.

These statistics, obtained from the U.S. Occupational Safety and Health Administration (OSHA) website, primarily address the regulatory hazards of health and safety management. Statistics relating to the health, productivity, and economic costs of unregulated issues, such as smoking, obesity, and workplace stress, are harder to compile but are undoubtedly just as disturbing and probably far more costly in the long run.

A well-designed and managed OHSAS HSMS can provide benefits beyond improved workplace health and safety. Some of these include:

- Improved morale—A safe and healthy workplace, where employees feel that the organization truly cares about their welfare, improves morale. In a study conducted by the University of Michigan, almost 87 percent of all workers interviewed considered that becoming ill or injured because of their job was at the top of their list of working condition concerns. Almost 81 percent were concerned about the physical dangers or unhealthy condition on the job.
- Reduced workers' compensation costs—Lower injury rates lowers workers' compensation costs. In addition, many insurance carriers provide a discount to organizations that have implemented a formal HSMS.
- Improved productivity—A cleaner and more organized workplace can significantly improve productivity. In this area some of the more popular quality initiatives directly support improved worker safety in addition to improved productivity and efficiency. One of the major benefits of lean manufacturing, and in particular 5S (a methodology focusing on maintaining a clean, orderly, and waste-free workplace), is in the identification and removal of workplace hazards.
- Improved market value—A company that has a solid safety record is valued higher in the marketplace than one that does not.
- Reduced financial costs—Lenders base the cost of capital for loaned funds on risk. A record of serious safety violations represents additional risks to the lender and will increase the interest rate proportional to that risk.

It is highly likely that many reading this book already have a formal HSMS in place or their organization has made a commitment to implement one. Why then spend time discussing the benefits of OHSAS 18001? An HSMS and its related policies, procedures, and practices are important and worthy of an organization's senior management team, supervisors, and rank-and-file employees support. Organizations may be able to minimally comply with safety regulations with a complacent and unengaged workforce, but it will not be able to go significantly **beyond compliance** without their enthusiastic buy in and ownership.

Management System Evolution

Management systems, like other systems, must evolve. Initially when an organization puts in a management system, the focus is on implementation and conformance with the new procedures and practices. Corrective action is taken when **nonconformance (NC)**, also known as an audit finding, is found. The focus is on getting a basic system in place. This is the *compliance phase*.

The benefits realized at this initial level of maturity are dependent on how well the informal practices that existed before implementation of the formal management system worked and how consistently they were followed. If a company had reasonably good management practices previously, then just "saying what we do" in the form of procedures is unlikely to significantly improve its health and safety performance (unless the procedures were not being consistently followed). These companies comprise a large share of the organizations that complain that they have not benefited from implementation of ISO 9001, ISO 14001, or OHSAS 18001. Organizations with good preexisting practices must rapidly move to the next level of maturity to benefit from their formal management system.

On the other hand, organizations without good preexisting systems or practices should realize significant improvements in health and safety in the compliance phase. The goal should be to move rapidly through the compliance phase toward the effectiveness phase. Unfortunately many companies never leave the compliance phase and after realizing initial benefits, complain that the management system no longer adds value. This phase is characterized by internal audits with numerous conformance deficiencies. For HSMS, this phase normally focuses on compliance with regulatory requirements and OSHA standards.

As the management system matures and conformance becomes less of an issue, more attention is focused on results. This is termed the *effectiveness phase*. In this phase both process owners and internal auditors focus more attention on the system and internal processes that make up the system producing results. In any large system with multiple interdependent processes, it is inherent that some processes or improvement objectives will not be producing the results that they should. The key to improving the overall system performance is to improve the results of each of the interdependent processes that feed into the system. In this phase, the management team and process owners will develop more robust systems of metrics to measure how well these internal processes are performing. Areas where results are not being obtained will be targeted for improvement. Internal auditors will spend more time looking at the effectiveness of processes and less at conformance. Internal audits will uncover more findings relating to effectiveness and **opportunities for improvement (OFI)**. NCs will be primarily associated with processes that are not performing adequately. Organizations in this phase will see improvements in their health and safety performance and bottom-line profitability through fewer lost work days, less absenteeism, and higher productivity.

The final level of maturity is the *continual improvement phase*. In this phase there is a complete system of metrics on all important processes, and these processes have been optimized to produce results. Although effectiveness and conformance will continue to be monitored, more attention will be placed on hazard elimination and error proofing. This is the level at which innovation, creativity, and the implementation of **best practices** becomes widespread throughout the organization. At this phase all of the primary benefits will be realized and with significant improvements in profitability. This evolutionary model of management system development is illustrated graphically in figure 1.1.

The bottom line is that the management system is what an organization makes it. If an organization is not realizing the expected benefits from a management system, it is time to ask what level of maturity the current system operates at and then take the steps needed to proceed to the next level. The OHSAS standard is a nonprescriptive minimal standard. It does not tell an organization exactly what to do nor how to do it. It

Figure 1.1   Management System Evolution

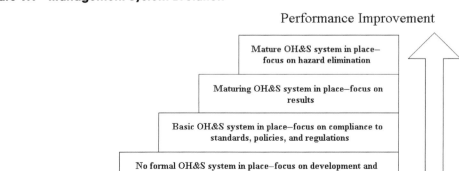

is minimal in that it provides the minimum requirements for health and safety management; it is up to an organization and its employees to go beyond the minimum in areas of importance.

The Elements of an Effective Health and Safety Program

The basic elements of a formal health and safety program include the following:

- management commitment and involvement,
- hazard identification, analysis, and control,
- hazard communication, and
- employee involvement and training.

MANAGEMENT COMMITMENT AND INVOLVEMENT

In theory, obtaining management commitment and involvement should not be difficult for the simple reason that by law management has a responsibility to provide a safe and healthy workplace for its employees. While they may delegate some of the responsibilities to others, the courts have shown that they cannot delegate their ultimate responsibility or accountability.

In practice, however, management commitment has often been overshadowed by the perceived costs (both in dollars and time) associated with safe work practices and by a general impression that most accidents happen because of worker errors and that workers should be responsible for their own safety. As a result, management involvement in many safety programs is minimal. In addition, there is a perception that adherence to the standards put forth and that form the basis for many regulations will result in a healthy and safe workplace. What is not recognized is that many of these standards were developed from consensus standards developed by the industry and as such represent minimum requirements that may or may not provide an adequate amount of protection in a given situation. While adherence to the practices and safeguards contained in these standards is necessary, they do not guarantee that no accidents will occur, especially when the human element is introduced.

An important concept to embrace is that *where no hazard exists, no accident can occur.* This properly puts the focus onto proactive hazard identification and elimination wherever practical and only then, on hazard control (and adherence to standards) where hazards cannot be eliminated. Focus on hazard elimination requires the full and active support of top management, because the mechanisms will involve multiple organizational functions, such as product and process design and development, purchasing, and finance, in addition to the historical involvement of production and maintenance. In addition, the steps needed to minimize the risks associated with hazards, where they cannot be eliminated, requires the consideration of the human element and will involve the creation of a safety-based culture throughout the organization, supported and driven by the company's management team.

OHSAS 18001 requires that top management demonstrate their commitment and involvement in the health and safety program in several ways, including the appointment of one of their own to serve as the management appointee. Additional discussion of the roles and responsibilities of the senior management team will be discussed in chapters 2 and 3.

HAZARD IDENTIFICATION, ANALYSIS, AND CONTROL

A necessary condition for establishing a safe and healthy workplace is the identification of the hazards that exist and the risks that they present under both routine and nonroutine operations. Too often **job hazard** analysis (JHA) is performed by untrained personnel, is done at an insufficient level, does not consider nonroutine operations, or is not done in a planned, systematic manner. In some companies JHA is an "after the fact" event that is conducted subsequent to an accident or injury. These do little to reduce worker injury inasmuch as they fail to identify all of the hazards that can lead to accidents or only focus on a single event.

The systematic and formal identification of workplace hazards, assessment of the risks presented by these hazards, and then identification and implementation of controls to minimize these risks lies at the heart of the OHSAS 18001 specification. Part 2 will examine methods that can be used to perform the JHA, risk assessments, and options for reducing or controlling their actual or potential effects in detail. The focus will always be on first trying to eliminate the hazard and then on safely controlling it if it cannot be eliminated.

HAZARD COMMUNICATION

Accidents and injuries can only be avoided if the hazards that lead to them are recognized, and the operational controls used to minimize their risks are understood and used. Compliance with OSHA's hazard communication requirements does ensure that information is available, but it falls short in ensuring this information is understood, available at all locations where it is needed, and that the controls used to protect the employee are effectively deployed. This is one of many areas where the organization needs to go well beyond compliance with the OSHA requirements to ensure worker safety.

In a more comprehensive sense, hazard communication also involves the establishment of systems to communicate policies, procedures, and best practices throughout the workforce and to obtain employee feedback regarding the existence of unrecognized hazards and the means to control them. It also involves the setting of conditions that lead to a total safety culture within the organization. OHSAS 18001's communication requirements include the development of a safety culture that includes involvement by employees in the setting of policies and procedures that lead to a healthy and safe workplace.

EMPLOYEE INVOLVEMENT AND TRAINING

Employees must be involved in the maintenance and improvement of the organization's health and safety program for it to be effective. Involvement generates ownership, and ownership drives both compliance and improvement. There are many ways to enlist the support and participation by the workforce; one of which includes training employees on the risks associated with their work, the consequences to themselves and to their coworkers of not following established policies and procedures, and on the methods that they can and must use to safeguard themselves and others. As noted by the University of Michigan study cited previously, most employees have a high degree of concern over their workplace safety and health and will freely comply with safety and health practices if management demonstrates their support for the health and safety program through its commitment of time, resources, policies, and funding. A system of disciplinary measures and peer pressure must be developed for those who do not willingly comply with organizational policy on health and safety matters. Midlevel managers and supervisors must be held accountable for safety performance within their areas of responsibility, and incentives should be provided for improved safety performance.

OHSAS 18001's requirements relating to employee involvement, training, and improvement address this component and provide a foundation for the implementation for additional programs, such as behavior-based safety systems.

Summary

The HSMS is but one of many components of the organization's business management system. All must function together as intended if the organization is to be successful in deploying its business plan and achieving its strategic goals. As more and more organizations trim staff, streamline operations, and deploy knowledge-based systems in response to global market pressures, the organization's remaining workers become even more important assets, and efforts to retain them and keep them productive becomes a strategic priority. The expense of replacing an injured worker or one who has left the company because of health or safety concerns can no longer be tolerated or simply replaced, given the intense competition businesses face. In addition, health care costs continue to rise faster than inflation, and many companies are now realizing that it makes good business sense to keep employees healthy and safe. Businesses need to go beyond minimal compliance with health and safety standards and regulations if they wish to become leaders within their industry. A formal HSMS based on OHSAS 18001 provides a proven model for ensuring the safety and health of the organization's employees who serve as the foundation for corporate success.

The next chapter takes a detailed look at the requirements of the OHSAS 18001:1999 specification. Each clause will be examined in turn, and explanations of the requirements provided. In addition, because many of the organizations considering implementation of 18001 already have an ISO 14001 environmental management system (EMS) in place, the requirements of OHSAS 18001 will be compared to those in ISO 14001 to highlight subtle yet important difference between the two models.

CHAPTER 2

Occupational Health and Safety Management Systems and the OHSAS 18001 Specification

OHSAS 18001 Basics

The OHSAS 18001 Occupational HSMS specification is used to assist in the deployment of an effective HSMS. The specification was modeled after the ISO 14001 EMS standard to provide ease of integration between these two mutually supportive documents. Because most countries in the developed world are required to maintain both environmental and health and safety management systems, and because the control of both of these programs share many common elements, it makes sense to have management systems that are as similar as possible. While the 2004 revisions to ISO 14001 made some changes such that not all of the clauses share the same numbering as they did before the revision, the essential requirements are still the same such that full alignment between the two documents is still relatively straightforward.

In 18001, the organization identifies the hazards associated with the workplace and workplace activities and then determines the risks that these hazards represent as part of the planning phase. This is accomplished through the systematic application of JHA methodology. Once these hazards and their associated risks are identified, the organization then implements appropriate controls to minimize the risks associated with these activities. It should be noted that a full JHA of all of an organization's hazards may take from several months to a year or more. Indeed, it is considered that the JHA process is a continual process owing to the constantly changing nature of workplace activities and processes.

The OHSAS 18001 HSMS specification provides a model for a comprehensive HSMS. The OHSAS 18001 specification, like ISO 14001, is developed around Dr. W. Edwards Deming's famous **Plan-Do-Check-Act (PDCA)** model of improvement. Its planning elements consist of the development of a health and safety policy that must include the organization's commitments to the health and safety of its employees. Of note, a commitment to comply with regulatory and other requirements is mandatory, as is a commitment to continually improve. As part of planning its occupational HSMS, the organization must also identify its workplace hazards. The organization then performs an assessment to determine which hazards could present a significant risk to its employees, contractors, or property. It then identifies controls and methods to eliminate those hazards or minimize their risks and impacts. Planned improvements, in the form of health and safety objectives and targets, must also be established along with plans on how to achieve them.

A subtle, but important difference between ISO 14001 and the occupational HSMS relates to the level of employee involvement. Both systems require a high level of organizational awareness and communication, but OHSAS 18001 goes much farther in that it requires a higher level of employee involvement in the development, maintenance, and improvement of the management system. This is termed *employee consultation* in OHSAS 18001.

The "doing" component is titled *implementation and operation*, and in this phase the organization implements the controls and safeguards identified in the planning phase. In addition, the specification requires that operators be trained in the safe and proper performance of their duties and be made aware of their roles and importance within the HSMS and the potential consequences of not following established safety policies and procedures.

The checking and corrective action elements represent the *check* component of the PDCA cycle. Here the organization monitors how well its health and safety controls are working and how its management system

11

is performing. Records of actions and performance must be maintained and controlled. Compliance to regulatory and other requirements must also be reviewed. If deficiencies or NCs are noted, then corrective action is initiated to restore the performance of the system.

The *act* component of the PDCA cycle is represented by the management review requirements, which require senior level review of the overall performance of the HSMS and its related components. The output of this review should lead to actions and decisions to correct or improve performance. The OHSAS 18001:1999 model is shown in figure 2.1.

In this chapter we examine each of the major clauses in the OHSAS 18001:1999 specification including the specific requirements and their intent. It will help if the reader has a copy of the specification handy as we walk through the requirements.

Introductory Information

The introductory material in the front of the OHSAS 18001 specification notes that OHSAS 18001 is not an international or British standard but rather is a specification to bridge the gap until one is developed. The International Organization for Standardization is currently working on such a standard, but the process is in

Figure 2.1   The OHSAS 18001:1999 Occupational Health and Safety Management System Model

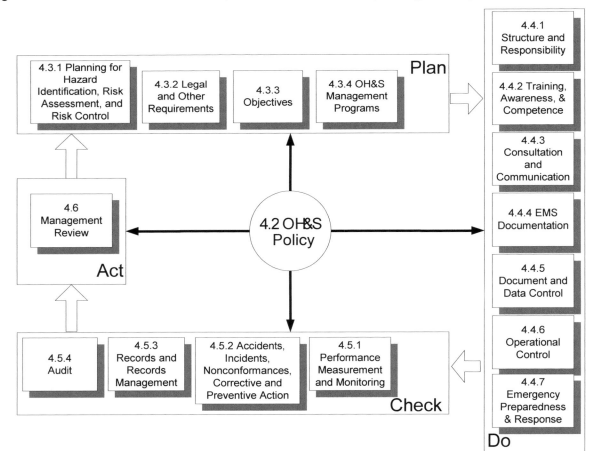

the early stage as of this writing and publication is years away. Once an international standard is issued OHSAS 18001 will be withdrawn. Issuance of a new international standard, when it does occur, should not require significant changes to the organization's HSMS, because the desire is to maintain a high level of compatibility with the ISO 14001 standard. In other words, any new international standard will look much like ISO 14001:2004 and hence OHSAS 18001.

SCOPE

Section 1 of the specification presents the scope. The scope defines the intended coverage of the specification. Of particular interest are subitems D through F that indicate how the specification can be used for demonstration of an organization's compliance. It notes that organizations can use the specification for certification, for second-party verification of an organization's compliance, or for self-declaration of compliance. This last option, self-declaration, allows an organization to implement and verify its conformance to OHSAS 18001 and then self-declare its conformity, thereby avoiding the costs of certification. While this option may be of interest to some companies, keep in mind that self-declarations normally will not satisfy any customers who require suppliers to be certified to OHSAS 18001. Currently, however, few customer organizations are requiring certification to OHSAS 18001, which makes this option more attractive.

REFERENCE PUBLICATIONS

Reference Publications relate to other documents that must be used in support of the specification. Note that for OHSAS 18001 there are two. Both OHSAS 18002:1999, *Guidelines for the implementation of OHSAS 18001* and BS 8800:1996, *Guide to occupational health and safety management systems* can be obtained from the British Standards Institute (BSI) and provide guidance for setting up a robust HSMS using OHSAS 18001 as the model. Neither contains requirements needed for certification.

TERMS AND DEFINITIONS

Section 3 of the specification contains terms and definitions that are used throughout the specification. Some of these are important. The most important of these are discussed here.

- Accident: The specification defines an accident as an undesirable event that leads to injury, death, ill health, or other loss. Other loss can include damage to property.
- Incident: An incident is the trigger event that leads to or has the potential to lead to an accident. The note adds that an incident with no effect is termed a *near miss*, which is an important (and often underreported) category of incident.
- Hazard: A hazard is defined as a source (e.g., a material, an emission, a machine, energy, etc.) or a situation (i.e., an activity or process that brings sources and opportunities for loss together) that has a potential for injury, death, ill health, or damage. A hazard can be considered to be the cause.
- Hazard identification: This is a process for identifying where hazards exist and determining the nature of the hazard. In the United States this is commonly termed JHA.
- Risk: The potential impact of a hazard, defined as the result of the consequences to personnel or property if the event occurred and its likelihood of occurrence. The risk can be considered the effect.
- Risk assessment: This is the process for identifying the magnitude of the risk presented by an event

and determining whether the risk is tolerable. The risk assessment follows the hazard identification (or JHA).

- Tolerable risk: The concept of tolerable risk has been defined as that level that can be endured by the organization as governed by its legal obligations and its occupational health and safety (OH&S) policy commitments.
- Objectives: OHSAS 18001 defines an objective as a goal set for OH&S performance. Note that OHSAS 18001 does not include the concept of targets as in ISO 14001. By not including targets there is less confusion over whether a performance requirement is a target or an objective.

Now that the definitions that apply have been examined, it is time to turn our attention to the requirements. The reader may want to refer to figure 2.1.

Main Clause 4.1 General Requirements

This clause basically says that the organization must develop a systematic HSMS that meets the requirements of the OHSAS 18001 specification.

Main Clause 4.2 Occupational Health and Safety Policy

This clause requires that the organization develop a health and safety policy. The policy statement is important. Everything that is done within the HSMS should be consistent and aligned with the policy statement. The specification requires that the policy statement:

- be defined by top management;
- be consistent with the organization's scale and nature of operations (i.e., OH&S risks found within the organization);
- be documented, implemented, and maintained;
- be periodically reviewed to ensure it remains relevant and consistent with the scale and nature of operations;
- be communicated throughout the organization; and
- be made available to the public (i.e., interested parties).

In addition to these basic requirements, the implementation of which will be discussed in part 3, the specification also requires certain content requirements for the policy statement. In particular, OHSAS 18001 requires the policy statement:

- include commitments to *at least* comply with current legal and other requirements applicable to the organization and
- include a commitment to continually improve its OH&S performance.

The term *other requirements* refers to voluntary requirements that the company has adopted, either through its own commitment to health and safety performance improvement or through implementation of customer requirements or corporate safety initiatives. Examples could include the OSHA Voluntary Protection Program (VPP). It could also include requirements or goals set forth in corporate safety policies or programs. It could also include customer-mandated requirements, such as those specified in the ISO/TS 16949 specification for automotive production part providers that requires that employee safety and health be considered during the design of new manufacturing processes.

Many companies only include the two mandatory commitments required by the specification in their policy statements. While this is acceptable, adding more specific objectives and commitments to the policy statement will allow for the setting of more focused organizational improvement objectives, which will be shown later.

Main Clause 4.3 Planning

CLAUSE 4.3.1 PLANNING FOR HAZARD IDENTIFICATION, RISK ASSESSMENT, AND RISK CONTROL

The focus in this specification is on the identification of what hazards need to be controlled and on the determination of the methods to be used to eliminate significant hazards where possible or on lowering their risk to tolerable levels. The specification requires that the organization maintain a process (not necessarily documented) to:

- identify workplace hazards and their corresponding risk on an ongoing basis and
- determine the methods that will be used to control these hazards and their associated risk. Control should be thought of as eliminating the hazard or minimizing the risk associated with it if it cannot be eliminated.

Note that the specification refers to the *ongoing* identification, assessment, and control of hazards and risks. This is an important concept, because this process is really never ending. With OHSAS 18001 the process of performing JHA and risk assessment on all major activities and throughout all parts of the organization can take years to complete and is never really finalized as a result of the ever-changing nature of the way an organization performs its tasks. Certification does not have to await completion of all JHAs throughout the organization. Rather, the establishment of the process and proof that it works through the conduct of a portion of the JHAs required serves as the benchmark that the organization is ready for a certification audit. A strategy for showing the organization's readiness for certification, in light of this never ending process, is presented in part 2.

A job hazard is simply a source or situation that can cause injury or illness to personnel or damage to property. A JHA is simply a systematic review of a task to see what types of hazards exist. As an example, think about a worker who is setting up a machine to run a part. As part of the task he or she must lift a twenty-pound cutting tool, place it properly on the machine bed, and securely fasten it in place. What types of hazards might be present? A few are listed here:

- the worker could receive a cut from the sharp metal surfaces of the tool;
- the worker could drop the cutting tools on his or her foot;
- the worker could strain or pull a muscle while lifting the twenty-pound tool;
- the worker could strike his or her hand or fingers while placing the tool on the machine or while fastening it in place (if the wrench slips); or
- the worker could be severely injured by operation of the machine if it is not properly secured before beginning the tool change out.

Note that this is not a detailed review of this activity and does not identify all of the hazards present during this task, but it does serve to illustrate the concept of a JHA. Once all realistic hazards have been identified for this operation the next step is to determine the risks of these hazards. Remember that risk involves two components, the consequence (or seriousness of the effect) and the likelihood of occurrence.

For the first hazard, we might determine that the consequence of a cut could be moderate, because the cut could cause an injury that would require medical attention. The likelihood of occurrence potential might be determined to be high, because there are numerous sharp edges on the tool; it is relatively heavy and hard to handle, and experience has shown that there have been numerous accidents resulting in cut hands (or near misses) reported on similar operations over the past two years. An overall assessment might deem this hazard then to be of moderate importance and one requiring formal controls.

Controls for this operation should include questions first: Can we eliminate the hazard through redesign? Probably not, because the sharp edges on the tool are needed for it to perform its function. Could we lighten the tool to make it less cumbersome? Again, probably not. Because the hazard cannot be eliminated we should then consider any engineering controls that could lower the potential for cuts. Maybe a simple edge guard could be fabricated to cover the sharpest edge on the tool during assembly (the guard would be removed once installed). This could have a secondary benefit of protecting the cutting surface from nicks or dings that could damage it during installation. As a final (or as a supplemental) option we might consider personal protective equipment (PPE), in this case providing the worker with appropriate gloves that would protect against cuts without increasing the likelihood of dropping the tool. Standardized work, a safety bulletin, or other administrative controls might also be considered.

The hazard identification and risk assessment would then be repeated for all hazards and all steps in this and the organization's other activities. When you start to think about how many tasks are performed in a typical organization you quickly see why the OHSAS 18001 specification calls for an ongoing hazard and risk identification process. The process used must:

- be defined in terms of its nature, scope, and timing;
- be proactive and not reactive;
- provide for the determination of risk categories to aid in assignment of operational controls;
- consistent with operational experience and the capabilities of the risk control measures employed;
- be used when developing facility requirements, training needs, and operational controls; and
- provide for the monitoring of actions taken both in terms of how effective they are and how timely they are implemented.

Note that the specification requires this process be applied to nonroutine as well as routine activities. Some of the most dangerous tasks are those that are not often performed. It also requires that all personnel, including subcontractors and visitors, and all facilities, which fall within the scope of the management system, be included in the assessments. Tools, strategies, and methods for organizing and conducting these reviews will be presented in part 3.

Finally, the specification says that the organization must document this information and keep it up to date. Facility modifications, process changes, material substitutions, and product engineering changes occur on a frequent basis. A process must be in place to determine which, if any, of these changes affect an employee's risk exposure. The specification also requires that these hazards and associated risks during the setting of OH& S objectives are considered.

CLAUSE 4.3.2 LEGAL AND OTHER REQUIREMENTS

OHSAS 18001 requires that the organization maintain a process to:

- identify the legal and other requirements that apply to the organization and
- to provide access to these requirements.

Essentially the requirements are directed at answering the following question—"*What requirements apply to us and where do I find them?*" Although the requirements look fairly straightforward, identifying the exact federal, state, or local regulatory health and safety requirements that an organization must meet can be difficult if the organization does not have health and safety specialists on staff. Many small- to medium-sized organizations simply limp along from year to year, complying with those regulations they are aware of and ignoring those that they are not. Be careful, however, because ignorance is no excuse in the world of regulatory compliance. Owners and operators can be cited and heavily fined for violations that they were not even aware existed. Establishing once and for all exactly what applies to the organization and what does not is the first step toward fulfilling the mandatory health and safety policy statement for compliance to legal and other requirements. A discussion on the term *other requirements* was included in the section on the OH&S policy previously discussed in this chapter.

The specifications also requires this information be kept up to date and be communicated to employees and other interested parties, such as subcontractors and visitors.

CLAUSE 4.3.3 OBJECTIVES

The organization must set documented health and safety objectives at relevant levels and functions of the organization. Objectives are major improvement initiatives. The specification states that these objectives must be consistent with its OH&S policy and notes that they should be measurable wherever practicable.

When setting these objectives, the specification also requires the organization to consider:

- legal and other requirements,
- financial, business, and operational requirements, and
- the views of interested parties.

Interested parties include employees and shareholders, the local community, families, regulatory agencies, or anyone else who has a stake in the organization's health and safety performance. Part 3 will discuss objectives in more detail and will provide methods to ensure interested parties are considered during the objective-setting process.

CLAUSE 4.3.4 OCCUPATIONAL HEALTH AND SAFETY MANAGEMENT PROGRAMS

Once the organization has decided on its objectives, it must develop management programs to ensure their achievement. Management programs are nothing more than action plans. The specification says that the programs must define who is responsible for achieving the objectives and targets, the means (and resources) needed, and the time frame by which they will be achieved, in essence, an action plan.

It also states that these management programs must be reviewed at planned and regular intervals and amended where needed to account for changes to the organization's activities, products, services, or operating conditions.

Main Clause 4.4 Implementation and Operation
CLAUSE 4.4.1 RESOURCES, ROLES, RESPONSIBILITY, AND AUTHORITY

Everyone must understand his or her roles and responsibilities for operation of the HSMS. The specification requires that roles, responsibilities, and authorities be:

- defined,
- documented, and
- communicated.

In addition, it requires that a management appointee be designated by top management. Roughly equivalent to ISO 14001's management representative, the management appointee's responsibilities include ensuring that the HSMS is:

- established,
- implemented, and
- maintained.

In addition, the management representative must also report to top management with recommendations for improvement. The primary difference between the management representative for ISO 14001 and the management appointee for OHSAS 18001 is the clear and direct requirement that the OH&S system representative must be a member of top management. Examples given in the specification include a member or the Board of Directors or executive committee. This is further reinforced by the wording that states that the responsibility for providing a safe and healthy workplace resides with top management. Designating the health and safety manager as the management appointee does not fulfill this requirement. It must be an officer of the company. Some organizations also designate a management representative (possibly the health and safety manager) who assists and reports directly to the management appointee.

Finally, this clause also addresses resources. Resources needed to implement and maintain an organization's HSMS must be identified and provided. Management demonstrates its commitment to the HSMS by providing the necessary resources. Resources could include human resources, infrastructure, technological resources, and financial resources.

CLAUSE 4.4.2 TRAINING, COMPETENCE, AND AWARENESS

OHSAS 18001 requires that all employees who could significantly impact health and safety be competent. Competency can be based on appropriate education, experience, or training.

The specification also requires that health and safety *awareness* training be provided to all employees. Organizations have traditionally been good at telling employees *what* to do and *how* to do it. This requirement mandates that an organization informs employees *why* they need to comply. People are much more inclined to support the system if they understand the health and safety consequences of not complying. Awareness training must include:

- importance of complying with policies, procedures, and the requirements of the HSMS;
- the significant health and safety benefits of improved personal performance;
- the OH&S risks associated with their work activities;
- their individual roles and responsibilities, including those associated with emergency preparedness and response;
- the potential consequences of failing to follow specified operating procedures.

The OHSAS 18001 specification also requires that training procedures take into account differing levels of responsibility, ability, literacy, and risk.

CLAUSE 4.4.3 CONSULTATION AND COMMUNICATION

An organization must develop procedures to communicate information relating to its HSMS. This system must include both internal communications and external communications with interested parties. External communications include government agencies, shareholders, suppliers, customers, contractors, and the local community.

This clause differs significantly from the ISO 14001 clause on communication, in that a much higher level of employee involvement in the design and maintenance of the HSMS is required. For ISO 14001 a small cross-functional team (CFT) may develop almost all of the organization's policies, procedures, and controls. Employee involvement could be limited to understanding and following company policies and procedures and maintaining an appropriate awareness of the significant environmental aspects and impacts of their work. For the HSMS, however, a higher level of employee representation and involvement is required as evidenced by the first two items noted in the following list. Employees must be:

- involved in the development and review of policies and procedures used to manage risks;
- consulted when there are changes that affect workplace health and safety;
- knowledgeable of who their OH&S representatives and who the management appointee(s) are; and
- must be represented on health and safety matters.

The means used to facilitate this employee involvement must be documented and communicated. Note that many organizations use their safety committee as the primary method for fulfilling these requirements. More discussion will be provided in part 3.

CLAUSE 4.4.4 DOCUMENTATION

This clause basically lays out the requirements for an OH&S policy manual. Although the policy manual is not specifically mentioned, nor therefore required, the specification does require electronic or hard copy of the information that provides an overall description of the main elements of the HSMS, how these elements interact, and reference to any documents that describe these activities in more detail. The easiest and most common method for providing this information is through a policy manual.

Somewhat like ISO 14001, the number of mandatory documented procedures is minimal. The organization must therefore carefully weight the advantages of documentation versus the disadvantages.

CLAUSE 4.4.5 CONTROL OF DOCUMENTS

The organization must control those documents that it determined it needed. Document control as required by OHSAS 18001 includes:

- review and approval of documents by authorized personnel,
- revising documents when needed,
- locating and making documents available where they are needed by users, and
- removing or otherwise ensuring that obsolete documents, including those kept for historical value, are identified as obsolete and otherwise safeguarded to prevent their unintentional use.

While almost every clause requires the development of a procedure (or process), procedures do not have to be documented. The organization must therefore carefully weigh the advantages of documentation versus the disadvantages. On the plus side, documentation helps to ensure consistency, provides a repository for operating criteria and other hard-to-remember information, and is a valuable aid when training new team members. Just as important, and often overlooked in written procedures, is that documentation also provides a place to describe nonroutine activities or how to respond to unusual situations. Finally, documented procedures form the basis for continual process improvement in that they describe and communicate the baseline set of actions that serve as the best current way of performing an activity. This allows others to study, challenge, improve, and subsequently communicate more efficient, effective ways of performing the task. This improvement or modification of processes can be accomplished in a controlled manner through the document control process to be discussed next, which prevents the infusion of ill-advised and possibly disastrous improvements.

The disadvantage of documentation is that you must develop, review, approve, issue, and control it. This requires some level of overhead effort. While some would argue that documentation also brings with it inflexibility, I would counter that this is the case only in situations where the documentation was not well written. In the final analysis, it is up to the organization to determine whether a documented procedure would add value or not. If so, develop a written procedure. If not, review and strengthen training and other controls and move on. It is a judgment call the organization must make. Internal auditors will help evaluate whether the company made the right decision as part of their management system audits as discussed in chapter 8.

4.4.6 OPERATIONAL CONTROL

Operational control consists of all those methods that will be used to control an organization's health and safety hazards and risks identified during the planning phase. The specification requires that the organization identify, plan, and control those operations and activities that are associated with its identified health and safety risks and are consistent with its policy and objectives. Establishing realistic, effective controls is an essential step toward improving an organization's health and safety performance.

Control means carrying out these activities under specified conditions, which includes:

- establishing and maintaining *documented* procedures in which their absence could lead to deviations from an organization's health and safety policy commitments or its ability to achieve its objectives;
- establishing and maintaining procedures for the design of facilities, equipment, processes, installations, and work organization to eliminate or reduce health and safety risks at their source;
- stipulating *operating criteria* in these procedures, where needed; and
- establishing, maintaining, and communicating relevant procedures and requirements to an organization's suppliers and contractors, where needed, to control the identified risks associated with procured products, services, and equipment.

Although the emphasis is on the development of documented procedures as needed, the organization will also use other controls to minimize any adverse health and safety impact of significant hazards. These may include maintenance practices, engineered safeguards, PPE, and administrative controls. The identification and deployment of appropriate operational controls will be discussed in more detail in part 3.

4.4.7 EMERGENCY PREPAREDNESS AND RESPONSE

An organization must identify potential accident and emergency situations and develop appropriate procedures to prevent or mitigate any injuries or illnesses that may result. Most companies already have plans and procedures to address emergency situations they may face as a result of the regulatory structure in the United States.

An organization should review these to ensure they include responses to any new potential situations that may have been identified as part of its initial JHA.

An organization's emergency preparedness and response procedures must be periodically reviewed and revised where necessary, including after incidents or emergencies. They must also be periodically tested where practicable. This is where most companies have weaknesses.

Main Clause 4.5 Checking and Corrective Action

4.5.1 MONITORING AND MEASUREMENT

In clause 4.4.6, an organization established operational controls to minimize the risks of workplace hazards. In this clause an organization ensures that its controls work. On a regular basis, an organization must monitor and measure key characteristics of its operations that could significantly impact health and safety.

Procedures for this monitoring must include:

- both qualitative (nonnumerical judgments) and quantitative (numerical) measures,
- methods to ensure monitoring of performance in achieving health and safety objectives,
- proactive measures of performance for conformance to the occupational HSMS, operational criteria, and legal requirements, and
- reactive measures of performance to monitor accidents, ill health, incidents (including near misses), and other indicators of weaknesses in the health and safety management system performance.

Monitoring and measurement must include data recording and analysis so as to be able to determine the need for corrective and preventive actions. The organization's monitoring and measurement equipment must be calibrated or verified and maintained. Monitoring equipment may include atmospheric sampling equipment, decibel meters, or other test equipment. Calibration and maintenance records must be retained.

Note that OHSAS 18001 does not have a dedicated clause for the monitoring of compliance to legal and other requirements similar to ISO 14001. Such monitoring is included in third bulleted item as noted previously.

4.5.2 ACCIDENTS, INCIDENTS, NONCONFORMANCE, AND CORRECTIVE AND PREVENTIVE ACTION

A HSMS system does not guarantee an organization will not have problems, but it must be able to react appropriately when these problems occur. OHSAS 18001 requires that the organization establish and maintain a process for dealing with actual and potential nonconformities and for taking corrective or preventive action. This process must provide for:

- investigation and handling of accidents, incidents, and nonconformance to requirements,
- appropriate action to mitigate the consequences that result from these incidents,
- the initiation and completion of corrective and preventive action to eliminate the cause, and
- verification of the effectiveness of these actions.

Note that OHSAS 18001 requires the organization to perform a risk assessment on all proposed corrective and preventive actions before implementation. This is to ensure that there are no unanticipated outcomes that could cause the cure to be worse than the original issue.

The specification also notes that the corrective and preventive action taken should be appropriate to the magnitude of the problem and the risk presented by the problems. The specification also requires that the organization update any documentation affected by the actions taken.

4.5.3 RECORDS AND RECORDS MANAGEMENT

The organization is required to develop processes to maintain records demonstrating that it is complying with the OHSAS 18001 specification, its local procedures and policies, and that it is achieving results. The specification makes it clear that audit results and reviews are included in these requirements. This process must address:

- the identification and traceability of records,
- records maintenance and storage,
- protection of records against damage, deterioration, and loss,
- record retrieval, and
- record retention and disposal.

Note that some records have regulated retention times and others have requirements for posting or communication. The organization must meet all regulatory requirements.

4.5.4 AUDIT

An internal audit program must be established. Audits must be conducted at planned intervals to determine whether the HSMS conforms to the OHSAS 18001 specification, has been properly implemented, is effective, and is being maintained.

The health and safety audit program must be based on the results of the organization's risk assessments and past audit results. The process must also ensure that auditors and audits are objective and impartial. The audit process must also address:

- auditor competencies,
- responsibilities and requirements for planning and conducting audits and reporting results, and
- the determination of audit criteria, scope, frequency, and methods.

Finally, the results of audits must be reported to management.

4.6 Management Review

Top management must perform periodic management reviews of the HSMS to ensure its continuing:

- suitability (does the system address the requirements of OHSAS 18001 and the organization's needs and policy commitments?);
- adequacy (are the resources sufficient to maintain and improve the HSMS?); and
- effectiveness (is the system getting results?).

To determine this, the specification requires that certain items be considered during the review. These include the need to make changes to the policy, objectives, or other elements of the HSMS based on internal audit results, changing circumstances, and the organization's commitment to continual improvement.

The focus of the management review is on monitoring and action. If an organization's management reviews never result in actions, then they are not effective. Note that the review must be documented and records retained.

Summary

This completes the review of the OHSAS 18001 HSMS specification. The purpose of this review is to equip the reader with an understanding of the requirements that must be met before launching into system design and deployment. To summarize this section, we should note that the OHSAS 18001 specification is broken up into five major sections or clauses. These five clauses can be recombined into five operational components as follows:

- A planning component: In OHSAS 18001 planning is represented by the initial and ongoing identification of workplace hazards and their associated risks, operational controls to minimize these risks, and its legal and other requirements. Planning for unexpected events is captured by the emergency preparedness and control requirements. Planning is centered on the establishment of an organizational policy that declares the company's commitments for health and safety performance.
- An operational component: In OHSAS 18001 requirements focused on deployment and maintenance controls specified during the planning phase are specified in the requirements for operational control.
- A monitoring and corrective action component: In OHSAS 18001 these activities are defined in monitoring and measurement, accidents, incidents, nonconformance, corrective and preventive action, and the HSMS audit clauses.
- An improvement process: Improvement is embedded in requirements for the setting of health and safety objectives and in the conduct of management reviews.
- Key support processes: These activities support the overall HSMS and include document and record control, structure and responsibility, communication, and training, awareness, and competence.

Part 2 focuses on the steps needed to develop an effective HSMS. Various strategies and tools will be presented that will allow the organization to do this in the fastest, most efficient manner possible.

Part 2

OCCUPATIONAL HEALTH AND SAFETY MANAGEMENT SYSTEM DESIGN AND DEPLOYMENT

In this part we focus on the design and deployment of an HSMS based on the OHSAS 18001 specification. The goal is to create a system that is simple, effective, and relatively easy to maintain. Emphasis will be on getting the system up and running quickly so that it can start providing benefits right away. By keeping the HSMS simple, administrative overhead will be reduced so that more attention can be spent on improvements that go beyond compliance. Tools and methods will be provided to help the organization meet these goals.

CHAPTER 3

The Planning Phase

In this chapter we focus on the initial steps required to develop an HSMS based on the OHSAS 18001 specification. Emphasis will be on the organizing and planning stage. One of the goals is to move quickly through this phase so as to maintain the momentum that should be generated after the initial decision to create a formal HSMS is made.

Project Overview

Like any management system deployment, the first steps involve getting organized and developing a project plan. The general sequence of steps for designing and rolling out the HSMS are shown in figure 3.1.

The time frames shown on the flowchart represent a general time frame for each phase, assuming a moderate level of effort is applied to the project. Adhering to this plan would result in a fully functional HSMS in less than twelve months. Unless an organization is unusually complex one should not extend the time for OHSAS 18001 implementation much beyond a year, because doing so often results in a loss of the momentum generated when the project is announced and the initial training is provided.

Please note that although each of these phases is shown in a series sequence, in reality many of the actions in the individual phases will be going on concurrently. These concurrent actions will be reflected in greater detail in the project plan.

The rest of this chapter will walk through each of these phases, step by step.

The Planning Phase

In the planning phase the initial steps needed to develop the HSMS are taken and the project is organized. A detailed flowchart for this phase is shown in figure 3.2.

MAKING THE INITIAL COMMITMENT

The first step in the planning phase is to obtain a commitment to develop and deploy a HSMS. Note that commitment is not the same as a decision. Commitment reflects a strong belief in the need for and benefits from good health and safety stewardship. While a decision to implement OHSAS 18001 can lead to a certified HSMS, commitment will be needed to produce the meaningful results envisioned by the specification's creators.

Figure 3.1   Flowchart for Development

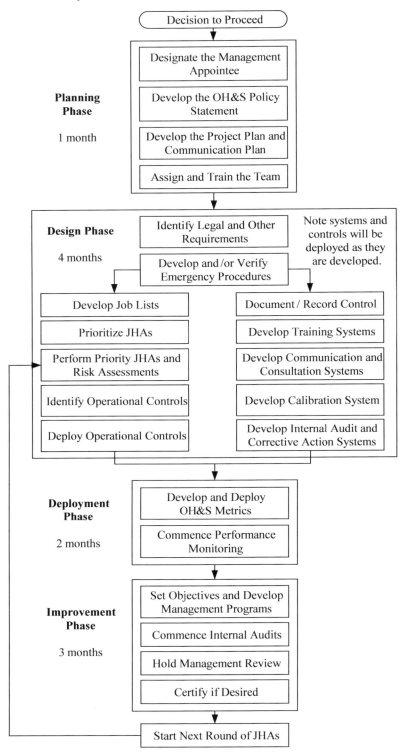

Figure 3.2 The Planning Phase

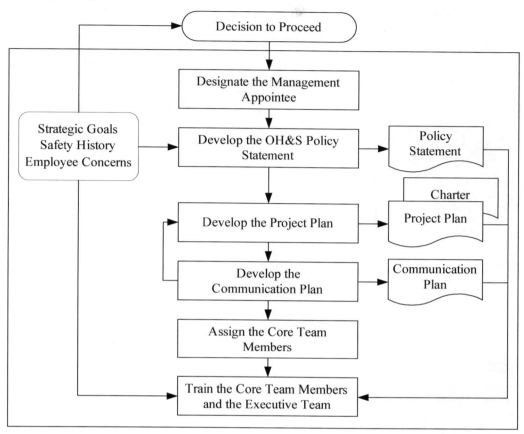

OBTAINING TOP MANAGEMENT AND ORGANIZATIONAL SUPPORT

The key to obtaining the commitment needed for an effective HSMS is to convince senior management that such a system is truly needed and can provide significant benefits for employees and the organization as a whole. The need should stem from more than just a customer requirement. The author recommends reviewing the benefits and statistics provided in chapter 1 with the management team, supplemented by historical company information. Case studies that illustrate the possible health, safety, and cost savings from implementation can also be shared.

The decision should be consistent with the organization's strategic goals, safety history, and employee concerns. Strategic reasons to implement a comprehensive HSMS could include increasing employee retention and morale, improving market image, or reducing health care costs. The organization's safety history may also be a factor, especially if the number or rate of workplace accidents is above industry averages. Finally, personal safety and health is a primary and fundamental human need, and an OHSAS 18001 implementation project can go a long way to addressing these needs, both practically and psychologically. Any of these reasons may serve as the basis for the decision to implement OHSAS 18001 and subsequently as a main driver for the changes that the system will bring.

Any discussion of top management support would be incomplete without a section detailing the role that senior management plays in success or failure of the management system. These responsibilities go far beyond providing resources in the form of personnel, time, and money to providing and reinforcing a culture that drives health and safety performance improvement throughout the organization. Some of management's most important responsibilities will now be discussed.

Providing Resources

The most fundamental of management's responsibilities is the provision of resources in the form of personnel, time, and training for the HSMS. Personnel resources include the assignment of CFT members and a management appointee. Perhaps even more important, management must be willing to provide the time for the CFT to develop the job lists, conduct the initial JHA and risk assessments, develop the systems and procedures, and design and implement the operational controls. A minimal CFT requires several individuals representing each operating area within the company. Money is also required to support the required training and the operational controls needed to reduce workplace risks. Even if the training is conducted internally there is an opportunity cost associated with the time away from work for the trainees.

Direction and Vision

Another responsibility of senior management is to provide vision and direction to the CFT. Senior management must be involved in the shaping and operation of the management program through the development and communication of a meaningful health and safety policy statement and must ensure that the HSMS quickly moves through the compliance phase into the effectiveness and improvement phases. Senior management must demand accountability for performance from process owners. Establishing a culture of performance and accountability will go a long way toward ensuring that all managers and supervisors take the management system seriously.

Communicating Commitment

It can be argued that the most important responsibility of senior management, however, is their active communication of the importance of worker health and safety and their support for the OHSAS 18001 management system. Communication means more than contributing a piece for the monthly company newsletter. Demonstrating the importance of the health and safety in daily actions and decisions is a much more powerful communicator of top management's support. Does management encourage research into improving the safety of the organization's products? Does management ask questions regarding the safety of new products during design reviews? Will management consider the use of new technologies to improve the health and safety of the workforce? Does management actively encourage and recognize suggestions on how to improve the company's health and safety performance and celebrate those that are implemented?

The OHSAS 18001 team leader and the top management will need to identify and communicate the drivers for change that resulted in a decision to implement a HSMS. Research has shown that many individuals will resist the change in practices, policies, and methods required to make meaningful improvements unless they can be shown why such changes are needed or unless they are provided some input into the ways things will be done in the future. Getting the organization's employees to embrace the new system is important, because many if not most of the more significant improvement opportunities come not from the senior management team but rather from frontline employees who know their processes the best. In this regard it is not unlike lean manufacturing, where the focus is on getting all employees involved in the identification and elimination of waste. For OHSAS 18001 the goal is to engage all employees in the identification of opportunities to eliminate workplace hazards or to minimize the risks of those that remain.

The development of a comprehensive communication plan will be discussed later in this section. This plan will include ideas and methods for obtaining organizational commitment to the OHSAS 18001 HSMS. Assuming that an organization has the top management team's commitment, it is now ready to proceed.

Designate the Management Appointee

As noted in chapter 1, a management appointee must be designated by senior management. While this individual will not do all the work required to design, implement, and maintain the HSMS, he or she will be ultimately responsible to see that it gets done. With that in mind, there are several considerations that should be taken into account before making this appointment:

- Does the individual have an adequate level of authority to make things happen? The OHSAS 18001 specification is quite clear that the management appointee must be a high level executive within the company. Examples cited in the specification are a member of the board or executive committee member. The HSMS will cross department boundaries. Will the management appointee have the authority to get the support needed to drive action when necessary? While the management appointee may have several others reporting to him or her regarding the operation of the HSMS, it is clear that ownership must reside at the top of the organization.
- Will this individual have the time needed to oversee the design, implementation, and maintenance of the system? The management appointee will need to devote quite a bit of his or her time to the HSMS, especially during the initial design stage. For busy executives consider the designation of one or more management representatives to help the management appointee with the day-to-day design, implementation, and improvement of the HSMS.
- Does the individual have the necessary knowledge? This consideration may be difficult to judge, especially in organizations without dedicated health and safety staff. The management appointee does not have to be an expert in health and safety engineering or regulations but should have the capability to learn about and understand health and safety topics as needed. In addition, familiarity with management system principles helps.
- Finally, does this individual have the desire and motivation to be the management appointee? Nothing can substitute for a belief in the job and a passionate desire to see the management system succeed.

CREATE THE CROSS FUNCTIONAL TEAM

The CFT can be assembled once the decision has been made to proceed. The team should consist of members of various functions within the organization but include as a minimum representatives from:

- engineering,
- facilities/maintenance,
- production,
- human resources,
- logistics,
- safety engineers (if the organization has dedicated staff), and
- safety committee team members.

In addition, representatives from the quality or environmental departments should be included if the organization already has a quality or environmental management system based on ISO 9001 or ISO 14001 in place to ensure compatibility with existing systems and procedures. The CFT will be responsible for organizing the initial JHA, development of system procedures and policies, and deployment of operational controls. They also often serve as internal auditors. Other employees may be brought onto the team as needed, but the CFT will serve as the permanent core team while the HSMS is being developed. Selection criteria for team members are similar to those set forth for the management representative, namely:

- interest and enthusiasm,
- good knowledge of the operations and activities within their departments and functions,
- availability (a commitment of four hours per week will be required for the first few months, along with a few eight-hour days for training), and
- good reputation and standing within the organization.

The last bullet is important as the CFT and the senior management staff will also serve as the initial ambassadors for the HSMS. They should be opinion leaders capable of generating enthusiasm to peers within their normal areas of responsibility. The management appointee will normally serve as the team leader.

As a final consideration it may be prudent to include one or more outside consultants to coach the team and provide expertise where lacking. As a minimum, it may be necessary to hire a consultant to help with the identification and documentation of regulatory requirements if this has not been previously developed and to perform the health and safety compliance reviews. These two tasks will be discussed in more detail later in this chapter.

Create a Project Plan

As with any project, the team leader should consider the development of a project plan to guide the implementation effort. The work breakdown structure will normally include the major phases of the project as outlined in this book, with tasks and deliverables outlined under each main phase. The project plan will help organize the project, assign resources, and otherwise ensure that the project stays on track. A portion of a project plan for a typical rollout is shown in figure 3.3.

In addition to the project plan, it is also often helpful to develop a project charter. This short summary document outlines the scope of the project, the time and effort involved, the goals of the project, and documents the management team approved to proceed. This is especially important for the team leader as it represents management's commitment to provide the resources needed to support successful project completion. An example of a project charter for an OHSAS 18001 implementation is shown in textbox 3.4. A full example of a project plan, project charter, and project scope statement are provided on the CD accompanying this book.

Provide Initial Training

Once the scope has been determined, the CFT assigned, and the project plan developed, it is time to initiate the team training. Training for OHSAS 18001 implementation is not extensive. A typical training series might include:

- One or more four-hour overview sessions for the senior management team, other managers, and supervisors.
- One two-day OHSAS 18001 training session for the CFT members and future HSMS auditors.
- Several two- to four-hour training sessions for personnel who will conducting the JHA and risk assessments.
- One three-day training session in HSMS internal auditing for HSMS auditors.
- Several short two-hour presentations to all employees introducing the OHSAS 18001 model, the company's health and safety policies, their roles, and other awareness training topics.

At this point only the four-hour executive overview sessions and the two-day CFT OHSAS 18001 training will be needed. The internal auditing training should be deferred until the system is essentially deployed and

Figure 3.3   Project Plan for OHSAS 18001 Implementation

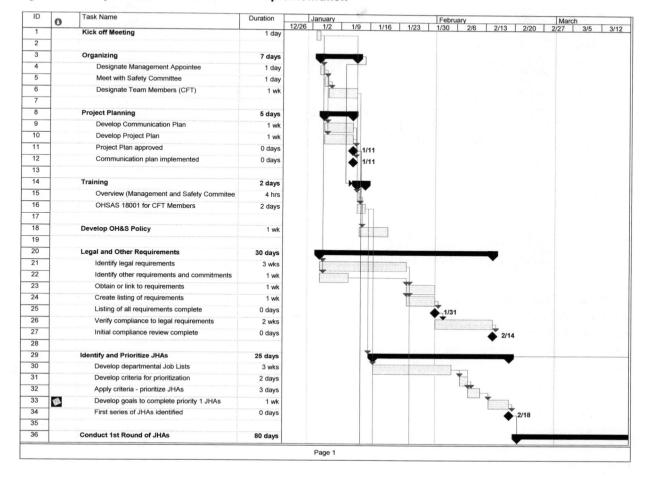

ID		Task Name	Duration
1		Kick off Meeting	1 day
2			
3		Organizing	7 days
4		Designate Management Appointee	1 day
5		Meet with Safety Committee	1 day
6		Designate Team Members (CFT)	1 wk
7			
8		Project Planning	5 days
9		Develop Communication Plan	1 wk
10		Develop Project Plan	1 wk
11		Project Plan approved	0 days
12		Communication plan implemented	0 days
13			
14		Training	2 days
15		Overview (Management and Safety Commitee	4 hrs
16		OHSAS 18001 for CFT Members	2 days
17			
18		Develop OH&S Policy	1 wk
19			
20		Legal and Other Requirements	30 days
21		Identify legal requirements	3 wks
22		Identify other requirements and commitments	1 wk
23		Obtain or link to requirements	1 wk
24		Create listing of requirements	1 wk
25		Listing of all requirements complete	0 days
26		Verify compliance to legal requirements	2 wks
27		Initial compliance review complete	0 days
28			
29		Identify and Prioritize JHAs	25 days
30		Develop departmental Job Lists	3 wks
31		Develop criteria for prioritization	2 days
32		Apply criteria - prioritize JHAs	3 days
33		Develop goals to complete priority 1 JHAs	1 wk
34		First series of JHAs identified	0 days
35			
36		Conduct 1st Round of JHAs	80 days

Page 1

audits are ready to begin. The two-hour all-hands presentations should wait until the system is ready to be deployed, so that essential information regarding key attributes of the system can be communicated.

Training for the executive team should focus on the importance of good health and safety stewardship, the benefits that can be expected from an effective HSMS, the basic structure of the system, an overview of the project plan, and their (i.e., management's) responsibilities for effective implementation, operation, and improvement of the system. A good idea is to follow this training up with a second two- to three-hour session a week later to develop the health and safety policy statement (see the next section).

Training for CFT members should include the topics provided during the executive overview in addition to a detailed review of the requirements contained in the OHSAS 18001 specification. I like to present the requirements in the order in which they will be implemented, but the flow and structure of the training is somewhat of a personal choice.

Draft the Policy Statement

The health and safety policy statement truly forms the heart of the HSMS, and the management appointee and CFT team leader must ensure that an appropriate amount of thought and energy is put into its develop-

Textbox 3.4   Project Chart

Project Charter

January 7, 2006

Project: OHSAS 18001 Health and Safety Management System Implementation

Business Need: A corporate decision has been made to implement an OHSAS 18001 health and safety management system at all WeProduceIt sites. This decision has been driven by the increasing costs of health care which have risen by 25% over the past 3 years. A formal health and safety management system will significantly lower both our insurance premiums and workman compensation costs.

In addition, health and safety concerns ranked high in importance in the last corporatewide employee survey. The OHSAS 18001 management system is thus seen as a tool to improve employee safety and satisfaction while contributing to a healthier and more productive workforce.

Expected Project Outcomes: Deployment and certification of an OHSAS 18001 health and safety management system that contains and/or reduces our health and safety costs.

Cost and Schedule: The project must be completed by December 15, 2006. The budget authorized for this project is $60,000, not including internal personnel time.

Personnel and Support: Chuck Bennet has been assigned as Project Manager and has been authorized to dedicate up to 25% of his time to this project through December 2006. Each department will provide support as needed and will assign a department lead to work with Chuck on this project. The primary Project Sponsor is Bill Clark, Chief Operating Officer and OH&S System Management Appointee.

_____	_____
President and CEO	VP. Human Resources
_____	_____
CFO	VP. Engineering
_____	_____
COO and Management Appointee	VP. Manufacturing

ment. Remember that the policy statement is a responsibility of the senior management team. It should not be developed in isolation by the management appointee or an outside consultant, rather it should stem from a careful debate of what the organization feels is truly important from an health and safety perspective. In essence, the health and safety policy statement is both a mission statement and a vision statement rolled into one. The commitments made in this policy statement will shape the rest of the HSMS and will go a long way toward determining whether an organization's system only achieves certification or accomplishes significant improvements in worker health and safety.

The requirements relating to the policy statement are found in clause 4.2 of the specification and were previously discussed in chapter 2. The art of developing an effective policy statement, however, rests with its ability to provide a framework for setting health and safety objectives. As discussed in chapter 2, this means going beyond the mandatory commitments of compliance and continual improvement. It requires the setting of specific commitments that are meaningful to the organization. By meaningful, we mean that the management team feels strongly enough about them so as to provide the time, resources, and financial funding to drive their improvement.

As an example, compare the two health and safety policy statements shown in textbox 3.5. Which one do you think will lead to meaningful performance improvement? The first policy statement (which includes the minimum mandatory commitments required by the specification) could serve as a framework for setting health and safety objectives, because the mandatory requirements are so broad and overarching as to include just about any improvement initiative launched. The problem is that it does not really focus the organization on anything.

The second policy statement provides more focus for setting objectives by noting a commitment to go beyond compliance where existing standards are shown to be inadequate in reducing risk to tolerable levels and by including a commitment to participate in OSHA's VPP. It also commits a focus on the long-term health and fitness of the organization's employees and to the establishment of a participative health and safety culture. Beyond the inclusion of the VPP in the policy commitments, which could in itself form the basis for a series of OH&S objectives, these policy commitments could drive the establishment of various employee wellness programs (e.g., smoking cessation, nutrition, and exercise and fitness programs) as a means to address the commitment to long-term health and fitness of employees and to the implementation of a behavior-based safety program in response to the commitment to provide a participative safety culture. Other objectives that could stem from this policy statement will be presented in part 3 of this book during the discussion of OH&S objectives.

If an organization is serious about implementing a HSMS that produces significant health and safety performance gains, then it should develop a policy statement that goes beyond the minimum commitments in OHSAS 18001.

Textbox 3.5   Health and Safety Policy Statements

WeProvideIt, Inc, is committed to the safety and health of its employees. In meeting this commitment, WPI will:

- Comply with all applicable health and safety laws and regulations.

- Comply with any other requirements relating to WPI's health and safety program.

- Strive for continuous improvement.

WeProvideIt, Inc, is committed to the continued safety and health of its employees. In meeting this commitment, WPI will:

- Continually work to identify existing job hazards and to eliminate or reduce their risks to our employees.

- Comply with all applicable health and safety laws and regulations, and to go beyond compliance where needed to safeguard personnel.

- Comply with any other requirements needed to protect our employees, including participation in OSHA's Voluntary Protection Program.

- Seek to continually improve its health and safety performance through programs that focus on the long-term health and fitness of our employees.

- Establish a safety culture that includes management, salaried, and hourly colleagues working together to jointly ensure the safety of all colleagues.

Figure 3.6 Organizational Change Process

```
          Identify the Drivers for
                 Change
                    ↓
          Organize and Plan the
                 Change

 Measure, Evaluate, and          Evaluate the Current State
        Improve
              Organizational Change
                     Process
     Implement the Plan            Develop the
                                Implementation Plan

          Communicate the Change
```

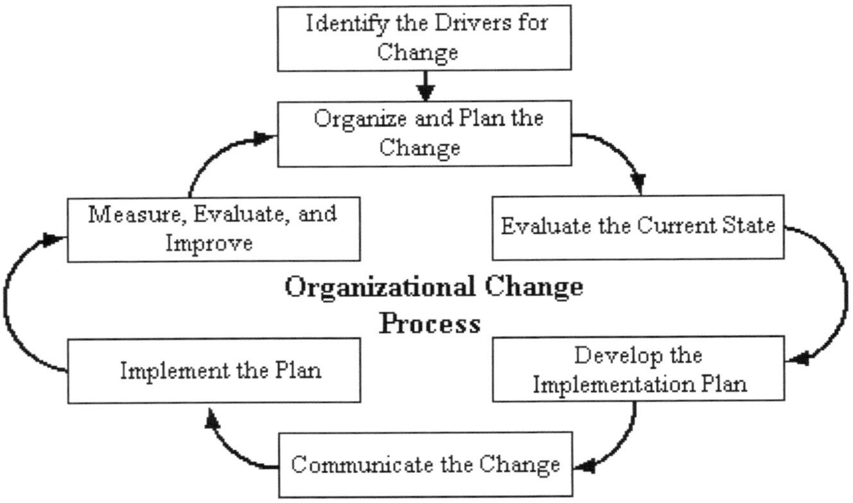

Develop a Communication Plan

Establishing an effective HSMS that goes beyond compliance will require significant cultural and operational change for most organizations, and change is never easy. A simplified view of the steps needed to succeed in leading organizational change is shown in figure 3.6.

The drivers for change were discussed in chapter 1. These could include a pressing need to drive down health care costs, to reduce current accident rates, or to respond to expressed employee concern. Note that in some industries, controlling or even reducing the long-term costs of health care is a survival issue. Both the organization and its employees are concerned with these rising costs; only in most organizations the concerns turn confrontational as management seeks to have employees pay a higher contribution to their insurance plans whereas workers seek to have the company continue to pay most if not all of the costs of their plans. The implementation of an OH&S management program, if properly designed, can go a long way to controlling these costs and can provide a mutual and common goal for both management and labor—preventing health care costs through accident reduction and the promotion of employee wellness.

The evaluation of the current state, organizing, and planning actions were discussed in the previous sections of this chapter. At this point, the focus is on communicating the change to everyone in the organization. An organization will also need to communicate with its insurers, suppliers, and contractors. The communication plan is the tool that an organization will use to target these communications to the various audiences. Some components of the communication plan are shown in figure 3.7.

These communication components serve as the basis for a change management plan and do not address all of the required communication requirements in the specification. Some of these additional requirements can be addressed now; others will have to wait until later in the HSMS development process. One of the communication requirements that can be addressed at this point is communication of the company's health and safety policy statement. As noted in chapter 2, everyone must be aware of the company's policy and should be able to relate it to their job activities. Methods used to accomplish this include the use of banners, plaques, inclusion of the policy statement in orientation training, and inclusion in some of the methods noted in figure 3.7, such as company newsletters and communication boards. Note that the policy must be made available to the public, and this is best accomplished on the company website and external communications.

Figure 3.7 Components of a Communication Plan

Component	Audience	Purpose	Methods
Initial Announcements (Top management team, CEO)	All employees	To generate awareness, involvement, enthusiasm, and support by defining drivers for change, how it will benefit the company and the employees.	All-hands meeting supplemented by company newsletter, formal letter, video, banners, and emails.
Periodic updates (Top management team, Management Appointee)	All employees	To create momentum by showing progress being made and contributions by other employees.	Articles in company newsletter, creation of communication boards, inclusion in periodic communication meetings, agenda topic during other management meetings.
Peer Leader Support	Safety Committee and/or Union Committee	To generate buy-in and ownership throughout the workplace by obtaining the commitment of peer leaders throughout the workplace.	Committee meetings, training, and cooperation in the development of the project plan.
Ongoing Awareness (All management and supervisors with employee contributions)	All employees	To increase involvement by recognizing and celebrating success throughout the workplace.	Permanent section in company newsletter, updating and expanding communication boards, and award ceremonies. Consider implementation of a company competition or award.
Supplier Notifications (Purchasing, engineering)	Suppliers, contractors, vendors	To communicate organization's commitment to the health and safety policies and procedures.	Purchase orders, contracts, supplier briefing statements, website, and formal letters.

Summary

At this point the decision to proceed has been confirmed and commitment obtained, a project plan developed, the management appointee and CFT assigned, the initial training completed, a health and safety policy statement defined, and a communication plan created. We are now ready to proceed into the HSMS design phase.

CHAPTER 4

Identifying Hazards, Risks, and Operational Controls

The critical steps in the design and development phase include the identification of the organization's legal and other requirements, planning for and identification of significant workplace hazards, identification of operational controls for the activities associated with these hazards, and development of the supporting systems and procedures needed to fully deploy the HSMS. These actions are shown on the flowchart in figure 4.1. This chapter will focus on the left-hand side of the design process (hazard identification and risk assessment), whereas chapter 5 will focus on the right-hand side of the flowchart (development of key supporting processes).

We will segment the design and development phase into three major activities:

1. identification of legal and other requirements,
2. development of a process to conduct JHA, risk assessment, and risk abatement and completion of the first round of JHAs, and
3. design and development of key supporting processes (chapter 5).

The reader should note that activities 2 and 3 should occur simultaneously. The planning and conduct of JHAs should be carried out by those with operational knowledge of the tasks or areas being evaluated, while the design and development of support system procedures will be completed by system experts. Also note that JHA should be an ongoing activity, not a one-time event.

Identifying Legal Requirements

For those organizations without dedicated health and safety specialists this may be the most difficult step in designing the management system. It is not unusual to find that an organization has a general feel for those regulations that apply to them but does not know with certainty which of the laws (federal, state, and local) they must meet, why they must meet them, and what they require. Instead, the organization continues to operate to what they think they must meet pending a state, federal, or local agency review and identification of weaknesses in their health and safety compliance. In many such cases the organization relies on an outside firm to conduct a compliance review every two or three years, but even then no one within the organization really completely understands what is needed to stay in compliance.

The problem with this mode of operation is that regulations change, as do the processes, facilities, materials, and methods used by the organization to produce their products or services. Without a solid knowledge of what the regulations are, the organization could find itself violating one or more of its regulatory obligations because of an increase in the amount or type of materials it uses or stores on-site, the implementation of new or modified processes, the acquisition and installation of new equipment, or modifications to its existing facili-

Figure 4.1   Process Flowchart for the System Design and Development Phase

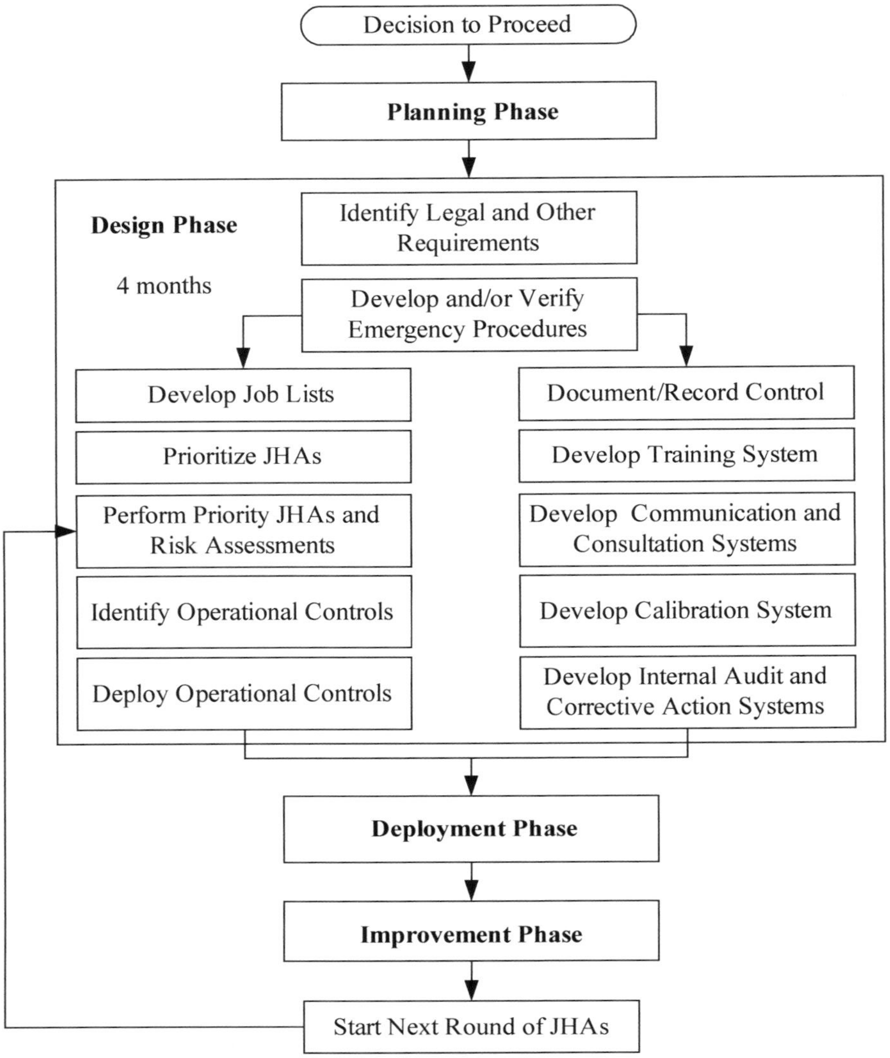

```
                        ┌─────────────────────────┐
                        │   Decision to Proceed    │
                        └─────────────────────────┘
                                    │
                                    ▼
                        ┌─────────────────────────┐
                        │     Planning Phase       │
                        └─────────────────────────┘
                                    │
                                    ▼
┌──────────────────────────────────────────────────────────────────────┐
│  Design Phase          ┌──────────────────────────┐                    │
│                        │  Identify Legal and Other │                    │
│                        │      Requirements         │                    │
│   4 months             └──────────────────────────┘                    │
│                        ┌──────────────────────────┐                    │
│                        │   Develop and/or Verify   │                    │
│                        │    Emergency Procedures   │                    │
│                        └──────────────────────────┘                    │
│                                                                        │
│  ┌──────────────────────────┐   ┌──────────────────────────┐          │
│  │    Develop Job Lists      │   │   Document/Record Control │          │
│  └──────────────────────────┘   └──────────────────────────┘          │
│  ┌──────────────────────────┐   ┌──────────────────────────┐          │
│  │     Prioritize JHAs       │   │  Develop Training System  │          │
│  └──────────────────────────┘   └──────────────────────────┘          │
│  ┌──────────────────────────┐   ┌──────────────────────────┐          │
│  │ Perform Priority JHAs and │   │  Develop Communication and│          │
│  │     Risk Assessments      │   │   Consultation Systems    │          │
│  └──────────────────────────┘   └──────────────────────────┘          │
│  ┌──────────────────────────┐   ┌──────────────────────────┐          │
│  │Identify Operational Controls│ │ Develop Calibration System│          │
│  └──────────────────────────┘   └──────────────────────────┘          │
│  ┌──────────────────────────┐   ┌──────────────────────────┐          │
│  │ Deploy Operational Controls│  │ Develop Internal Audit and│          │
│  └──────────────────────────┘   │  Corrective Action Systems │          │
│                                  └──────────────────────────┘          │
└──────────────────────────────────────────────────────────────────────┘
                                    │
                                    ▼
                        ┌─────────────────────────┐
                        │    Deployment Phase      │
                        └─────────────────────────┘
                                    │
                                    ▼
                        ┌─────────────────────────┐
                        │    Improvement Phase     │
                        └─────────────────────────┘
                                    │
                                    ▼
                        ┌─────────────────────────┐
                        │  Start Next Round of JHAs │
                        └─────────────────────────┘
```

ties. It is important to remember that the owner and operator of the facility, and not the outside consultant, bear the primary responsibility for compliance.

For an organization to meet the OHSAS 18001 specification requirements regarding legal requirements, the company must identify all of its regulatory obligations, ensure it has access to what the regulations or standards say, and understand what they have to do to meet these requirements. A partial listing of some of the health and safety standards that may apply include:

- General Safety and Health Provisions (Subpart C)
- Walking and Working Surfaces (Subpart D)
- Means of Egress (Subpart E)
- Powered Platforms, Manlifts, and Vehicle-Mounted Work Platforms (Subpart F)
- Occupational Health and Environmental Control (Subpart G)
- Hazardous Materials (Subpart H)

- PPE (Subpart I)
- General Environmental Control (Subpart J)
- Medical and First Aid (Subpart K)
- Fire Protection (Subpart L)
- Compressed-Gas and Compressed-Air Equipment (Subpart M)
- Material Handling and Storage (Subpart N)
- Machinery and Machine Guarding (Subpart O)
- Hand- and Portable-Tools and Other Hand-Held Equipment (Subpart P)
- Welding, Cutting, and Brazing (Subpart Q)
- Special Industries (Subpart R)
- Electrical (Subpart S)
- Commercial Diving Operations (Subpart T)
- Toxic and Hazardous Substances (Subpart Z)

These regulations appear in title 29 of the Code of Federal Regulations (CFR), part 1910 and are available online through various government websites. The parenthetical references refer to the subpart within the code where the requirements will be found. They can also be purchased in hard-copy format.

States also have their own regulations covering roughly the same topics as do the federal laws. Many states have the authority to implement and administer the federal regulations. In these instances the state law must be at least as restrictive, and often is more so, than the federal regulation; so if the organization complies with the state laws then it is also in compliance with the federal law. In cases where the state does not have the authority to administer the federal regulation, then the organization must stay aware of both the state and the federal regulations to ensure its compliance. In addition, most municipalities have their own local regulations or codes regarding fire safety, building safety, and emergency response.

To keep all this information straight, an organization needs to develop a listing, or register as it is often called, of its regulatory obligations. This is easier said than done, because the health and safety codes, regulations, and standards are themselves complex and can be hard to use for those without specific training in their use. For these reasons, it is recommended that a health and safety regulatory expert create the register of regulations. If the organization has dedicated health and safety professionals they should be able to create the register in short order. If the organization does not have such personnel on-site but is a member of a larger corporation it is often possible to get corporate specialists to develop the list. If neither of these two options are possible then it is best to hire an outside expert to create the list.

Resources are also available from various government agencies to help the organization determine which laws or standards apply to them. An example from the state of Michigan is shown in figure 4.2 and was used in conjunction with Michigan Manufacturers' Guide to Environmental, Safety, and Health Regulations. Although the checklist does not provide a complete, comprehensive guide to all of the regulatory requirements that an organization may have to meet, it is a good starting point for determining an organization's legal obligations. The numbered citations under the "Yes" box refer to the sections in the guide that discuss the applicable state requirements.

The most important thing is that the list must have all the information needed by the organization to understand its regulatory obligations and to access them when needed. For this reason I recommend that a format similar to the one shown in figure 4.3 be used, as a minimum, when creating the listing. If an outside consultant is used, he or she should be shown the required format before he or she conducts his or her review to ensure it will meet an organization's needs in satisfying the OHSAS 18001 specification requirements and helping an organization stay in compliance in between outside consultant reviews.

Once the register is developed, an organization should establish hyperlinks to the online versions of the state or federal regulations. The CFR can be accessed at the Government Printing Office website at www.gpo.gov.

Figure 4.2   Example Checklist for Determining Applicability of Michigan Health and Safety Laws

SECTION TWO—MICHIGAN STATE REGULATIONS

Part 1: Common Regulations for Safety and Health

60. Chapters 13, 14 and 15 deal with subjects that are applicable to all manufacturers. It is recommended that you read these chapters.

61. Are there any hazards at your facility that you are unable to eliminate through safeguarding or engineering changes and require the use of personal protective equipment? (*e.g., safety goggles, respirators, gloves, etc.*)
  ☐ Yes—Chapter 16
  ☐ No—continue

62. Do any of your processes involve toxic, reactive, or flammable chemicals?
  ☐ Yes—Chapter 17
  ☐ No—continue

63. Do you manufacture, use, store, sell, or transport explosives, blasting agents, or pyrotechnics? (*Excluding the sale or use of public display pyrotechnics (fireworks)*)
  ☐ Yes—Chapter 17
  ☐ No—continue

64. Do any of your processes or operations require that employees enter a confined space? (*e.g., tanks, hoppers, storage bins, vaults, etc.*)
  ☐ Yes—Chapter 18
  ☐ No—continue

65. Does your operation involve spray finishing?
  ☐ Yes—Chapter 19
  ☐ No—continue

Part 2: Health Regulations

66. Are any employees at your facility exposed to hazardous chemical fumes, vapors, or any other air emissions within the facilty?
  ☐ Yes—Chapter 20
  ☐ No—continue

67. Do any employees at your facility work with asbestos-containing materials?
  ☐ Yes—Chapter 21
  ☐ No—continue

68. Do any employees have the potential to be exposed to blood or other potentially infectious materials (OPIM)? (*OPIMs include semen, vaginal secretions, and several internal body fluids. OPIMs do not include sweat, tears, saliva, urine, feces, and vomitus, unless they contain visible blood or OPIM.*)
  ☐ Yes—Chapter 22
  ☐ No—continue

69. Does your facility store hazardous substances in quantities that could require an emergency response if released? (*An emergency response is a response effort by employees from outside the immediate release area or by other designated responders.*)
  ☐ Yes—Chapter 23
  ☐ No—continue

70. Are medical services (clinic, ambulance, hospital, etc.) "readily accessible" at your facility? (*Readily accessible means within ten minutes travel time*)
  ☐ Yes—Chapter 24
  ☐ No—continue

Most states have their regulations similarly posted online. Establishing hyperlinks makes it easier to access the regulations that apply to an organization and simplifies document control, because the government agency will keep these regulations up to date. One word of advice, however; look through the online documents after the hyperlinks are established. An organization will need to be familiar with both the regulations and with navigating through them. It does not have to be a safety expert, but the person should know what applies to an organization and should have a good working knowledge of the requirements.

Figure 4.3 Register of Occupational Health and Safety Regulations

REGISTER OF REGULATIONS FOR WEPROVIDEIT, INC.

Item No.	Regulation	Applies?	Rationale	Status	Actions	Location of Regulation
1.1	**Hazard Communication** 29CFR 1910.1200 and MI PA 154	Yes	Exposure to hazardous chemicals is possible.	Substantially compliant	Maintain compliance with Michigan PA 154	http://www.access.gpo.gov/cgibin/cfrassemble.cgi?title=200 http://www.michigan.gov/deq/0,1607,7-135-3307_4132
1.2	**MIOSHA Posting** Michigan Administrative Rule Part 13	Yes	All Michigan employers	Substantially compliant	Continue to maintain posting in accordance with Michigan Administrative Rule part 13	http://www.michigan.gov/deq/0,1607,7-135-3307_413214902—,00.html
1.3	**Record Keeping** Michigan Admistrative Rules Part 11	Yes	Employ more than 10 employees	Substantially compliant	Continue to maintain MIOSHA 300 and 301 logs/records	http://www.state.mi.us/orr/emi/admincode.asp?AdminCode=Single&Admin_Num=33601201&Dpt=EQ&RngHigh=
1.4	**Personal Protective Equipment** MIOSHA General Industry Standards Part 33 and General Industry Occupational Health Standards Part 433	Yes	Workplace hazards exist that require the use of personal protective equipment	Substantially compliant	Continue to comply with MIOSHA standards	http://www.access.gpo.gov/nara/cfr/waisidx_05/40cfr122_05.html http://www.michigan.gov/deq/0,1607,7-135-3307_413214902—,00.html#Waste

Other Requirements

Other requirements include customer requirements, industry requirements, corporate commitments, and voluntary participation in local, state, federal, and international programs that focus on health and safety.

An example of a voluntary initiative includes participation in OSHA's VPP or an organization's state equivalent. These programs are voluntary, and participation is not a requirement for OHSAS 18001 certification. If an organization is committed to improving its health and safety performance beyond compliance, however, these programs should be seriously considered. These programs provide guidance, technical assistance, and recognition beyond what can be provided by simple compliance, and they can also help unite an organization toward a common goal.

Once an organization has made a commitment to participate in one or more voluntary programs then OHSAS 18001 expects that it will establish methods, controls, and measures to ensure that the expected performance is achieved. An organization should include these programs or links to the websites that describe these programs in their listing of requirements. More information on OSHA's VPP and how OHSAS 18001 can be used to support it is provided in chapter 8.

Finally remember that not all workplace hazards are specifically addressed by regulations and standards. Even so, the organization is still required to provide a safe and healthy workplace under the OSHA's General Duty Clause which states "*Each employer shall furnish to each of his employees employment and a place of employment which is free from recognized hazards that are causing or are likely to cause death or serious physical harm to his employees.*" Indeed, the prime objective of JHA is to identify where workplace hazards exist so that the employer can fulfill his or her legal obligation in accordance with the General Duty Clause.

Reviewing and Developing Emergency Preparedness and Response Procedures

Another action that should be taken during the design phase is to review and where necessary modify (or develop) emergency preparedness and response procedures. These would include the:

- fire and general emergency plan,
- evacuation plan,
- spill prevention, countermeasures, and controls plan (SPCC),
- spill response procedures, and
- any other site-specific plans unique to the company, industry, or state in which the company operates.

In most western countries, including the United States, there are regulatory requirements associated with these plans and procedures. For the most part, therefore, these procedures already exist. At this point the team should review these procedures and plans to verify how current they are, their content, and their use in addressing any new serious accidents that may not have been recognized when the plans were developed.

As an example, one company I worked with identified the existence of a number of mercury switches in their production machinery during their initial environmental review for an ISO 14001 management system. They planned to replace these switches but knew it would take some time to do so. They already had a general spill procedure but that procedure only addressed spills of solvents and oil. They subsequently modified this procedure to include actions in the event of breakage of one of these mercury switches. The information on how to appropriately respond to this type of event was available on the state department of environmental quality website, but it required some searching to find it. Subsequently a mercury switch was in fact broken resulting in a spill of solid mercury. Because the company had already recognized the potential for such an event and modified their spill procedure to include the appropriate response, the spill was handled quickly and safely without the need for outside resources.

Also keep in mind the title of this clause in OHSAS 18001—emergency *preparedness* and response. In my experience most procedures and plans are heavy on the response and light on the preparedness. The design of an organization's HSMS provides them with an opportunity to review and possibly improve the prevention and preparedness portion of its plans and procedures.

Finally, an organization should make sure that its procedures and plans reflect current contacts, responsibilities, and phone numbers. It is almost a given that some of this information will be out of date. I have audited many a company whose emergency response coordinator no longer works at the company. These plans must be regularly reviewed and updated.

Job Hazard Analysis and Risk Assessment

The heart of the occupational HSMS is the JHA and risk assessment. During this activity the organization will evaluate its operations and facilities to identify workplace hazards. Once it has identified these it will conduct a risk assessment to determine which of these tasks require additional operational controls to safeguard personnel, plant, and equipment. Finally the additional controls needed shall be identified, developed, acquired, and then implemented as part of the next phase, implementation. Unlike an ISO 14001, initial environmental review, which can be accomplished in a matter of days, the completion of a complete set of JHA can take months or even years and thus requires careful planning and preparation. This section will provide guidance on how to prepare for the hazard analysis and assessment and provides tools for conducting the reviews.

SELECTING THE JOB HAZARD ANALYSIS TEAM MEMBERS

One of the first steps in preparing for JHA is to select and train the individuals who will be performing the JHAs. The team should be multidisciplinary, including members from each functional department or operating area of the facility. Some options for selecting individuals are detailed here. An important assumption, no matter which option is chosen, is that those assigned to perform JHAs will be given adequate time and the specific training required to complete the analysis.

- *Use safety engineers.* This option is desirable if the organization has trained safety professionals on staff. Safety engineers can be counted on to do a thorough JHA. Unfortunately this option does not build the capability to recognize and respond to workplace hazards throughout the workplace. As an alternative, consider pairing up safety engineers with members of the CFT, safety committee, or area supervisors to conduct the JHAs. This will provide the dual benefits of ensuring the quality of the JHA while expanding the ability of the organization to recognize and respond to workplace hazards.
- *Using members of the CFT* formed during the initial design phase. Advantages to using this core group of individuals include better control of the JHA that is conducted and a preexisting awareness of the OHSAS 18001 specification and the importance of its effective implementation. Constraints arguing against use of the CFT include incomplete knowledge of the large number of tasks that need to be evaluated and the need for the CFT to work on designing the system supporting processes during this same time period. It may also reinforce an apparent transfer of responsibility for workplace safety from the area supervisor to the CFT members performing the analysis.
- *Use of the plant safety committee.* Assuming that the plant safety committee includes members from each area of the plant, then this can be an attractive option. Advantages include the existence of a wide variety of skills and operating experience, strong synergy between the formal responsibility of the safety committee and the activities associated with the JHA, and a strong interest on the part of the committee members to improve health and safety. In addition, most safety committees include some management representation, which can be useful when the time comes to allocate funding for needed operational controls. Disadvantages may include the feeling on the part of some members that JHA is outside the scope of their responsibilities, the presence of a particularly loud or dominant manager on the committee, and the belief that the performance of the JHAs, risk assessments, and associated reporting detracts from the time available to fulfill their other responsibilities.
- *Use of area supervisors and team leaders.* A primary argument for using area supervisors and team leaders is that these individuals already have responsibility for the health and safety of the employees that report to them, and therefore the analysis of job hazards within their respective areas should already be a normal part of their job. Training them in hazard recognition and assessment provides the knowledge to properly carry out this responsibility. Additionally, these individuals have a good working knowledge of the tasks that are performed in their areas, have the authority to enlist others within the department to assist where needed, and often have more discretionary time than do others to complete the JHA. The main disadvantage for this option is that some supervisors or managers may view health and safety as an employee responsibility and may therefore ignore or overlook workplace hazards that should be addressed.

The last option is the best long-term option, because it provides the frontline supervisor with not only the responsibility but also the tools and knowledge needed to ensure the health and safety for those who report to him or her. It can also simplify the scheduling of JHAs, because the supervisor is already assigned to the area and therefore has an easier time fitting in the JHA between other routine duties. Finally, it helps to disseminate the skills needed to recognize and respond to workplace hazards throughout the workplace. Note that it does require the presence of a strong safety culture that extends from the frontline employees though all levels of management.

TRAINING THE JOB HAZARD ANALYSIS TEAM

Only those trained in hazard recognition should be authorized to perform JHA. Training can range from a few hours to a full day, depending on the depth of the analysis that will be accomplished and the work experience of those being trained. In particular, training in musculoskeletal disorders (MSDs; also commonly called *ergonomics*) can add several days to the training curriculum but can be valuable as ergonomic injuries represent one of the fastest growing categories of lost workdays and workman compensation claims. Job safety analysis training should also be made part of mandatory training for the new supervisor in organizations where the responsibility for JHA completion rests with the frontline supervisor.

Training should consist of an orientation to process mapping, general types of workplace hazards (e.g., struck by, caught between, exposure to, etc.), risk assessment, and the main categories of operational controls that can be used to reduce risks. Trainees should include the actual performance of a JHA and risk assessment while being observed by the instructor. Case studies of accidents that have occurred in the plant or in similar plants should also be used. Finally, the training should include a short discussion of interviewing techniques to ensure that those being observed will feel comfortable when their tasks are being analyzed.

Before the training it is recommended that some of the tools that will be used to conduct the analysis be developed so that these can be introduced during the training. At a minimum, the following tools will be needed during the review:

- a checklist to help guide the team in what to look for,
- the criteria that will be used to asses risk, and
- process flowcharts, if available, for major production processes.

Two examples of tools for helping to identify workplace hazards are provided in figures 4.4 and 4.5. The first shows a worksheet suitable for analyzing a manufacturing or service process, where hazards are normally identified through the listing of all actions needed to perform the process. The second checklist provides helps in identifying hazards found in general areas, such as administrative and engineering spaces, warehouses, facilities, and maintenance areas. Both should be used during the review. Copies of the tools are provided on the CD accompanying this book.

JHA of specific tasks would use a form similar to figure 4.4. There may be twenty, thirty, or more JHAs performed in a specific area, depending on how the processes were broken down. There would only be one area hazard review. The purpose of the area hazard review shown in figure 4.5 is to identify those workplace hazards that may not be directly associated with any of the organization's primary tasks. These types of hazards include the condition of stairs, walkways, aisles, the storage of hazardous materials, the presence of electrical switchgear, and the operation of auxiliaries, such as air and hydraulic systems. The combined use of JHA and area hazard identification ensures that no significant workplace hazards escape our scrutiny.

The use of these tools will be discussed in more detail in the next section. For now, recognize that the training of the JHA team members should include training on the use of the tools that they will use when performing the hazard analysis and risk assessments.

Figure 4.4   Job Hazard Analysis Worksheet (Process)

Plant:	Department:					
	Sub-Department:		JSA No.			
Job/Task:	Page: 1 of 1			Updated:	Draft ☐ Final ☐ Revised ☐	
Position(s) Performing The Job/Task:	Supervisor(s):			JSA Performed by:		

Required Equipment for the job:	
Job specific required P.P.E.:	
Job specific optional P.P.E.:	

	Job Procedures- Break down the job into basic steps, chronologically. This is done by observation, discussion with the operator and personal knowledge. List each step that has an observable hazard. Include routine and non-routine actions.	Potential Hazards- Barring all protective measures, both in place and potential, what could potentially happen to the operator performing this job? (A Potential Hazard list is found below.)	Risk / Hazard Class- Using the given matrix, what is the severity and likelihood rating of the potential accidents?				Action/Procedure to Control/Eliminate- What protective measures are in place or can be taken to eliminate and or minimize the risk of injury at this job?
			Severity	Likelihood	Total	Category	
			1-3-9	1-3-9	1-81		
1							
2							
3							
4							
5							
6							
	Overall Rating (highest individual ranking):						

Potential Hazards: SB-Struck By, CW-Contact With, Cby-Contact By, CB-Caught Between, SA-Struck Against, CI-Caught In, CO-Caught On, O-Overexertion or Repetitive Motion, FS-Fall at Same Level, FA-Fall From Above, E-Exposure to Chemicals, Vapor or Noise

Severity Rankings: 1 (Slightly Harmful) - superficial injuries, dust in eyes, minor cuts.   **3** (Harmful) - cuts requiring treatment, burns, fractures, sprains, dermatistis, concusions.   **9** (Extremely Harmful) - loss of limb, major fractures, disabling inju

Liklihood Rankings: 1 (Extremely Unlikely) - Never happened before, muliple levels of control, extremely low industry occurences.   **3** (Unlikely) - Rare occurrence in past or near misses, at least one safeguard in place.   **9** (Likely) - History of occuren

Risk / Hazard Category: 1- Neglible (N)   **3-** Tolerable (T)   **9-** Moderate (M)   **27-** High (H)   **81-** Unacceptable (U - Do not operate until risk is reduced):

In addition to the checklists, the management appointee or project leader should obtain a facility layout drawing that shows the general areas of the plant. This layout will be used to identify areas to be evaluated and to make team assignments.

JOB HAZARD ANALYSIS AND RISK ASSESSMENT

A diagram of the JHA and risk assessment process is shown in figure 4.6. We will now walk through this process prior to reviewing in detail the methods that can be used.

1. *Hazard Identification.* In this first step, the individual tasks performed by the organization will be analyzed using JHA, **failure mode and effects analysis (FMEA)**, fault tree analysis, hazard and operability study (HAZOP), or other formal technique. In this book all of these techniques will be referred to as JHA. The purpose is to identify sources of hazards. Each source (jobs, areas, etc.) may and usually does have multiple hazards. The inputs on the left refer to tools used to conduct the hazard analysis, whereas the inputs entering the top represent resources that can be used to prioritize the tasks that need analysis. It is important to realize that sources can be associated with the actions required to perform a task or with the materials and equipment located in the general area. Finally, note that both routine and nonroutine activities need

Personal Declaration: I declare that I have been advised of the job related hazards associated with the operation of my duties described in this JSA and have received training on the operations. I also acknowledcge that I have receive d the required PPE f			
Printed Name and Clock Card No.	**Signature**	**Date**	

to be analyzed. In many cases nonroutine activities present more of a hazard than do routine activities resulting from an absence of operational controls or worker unfamiliarity with the steps involved.

2. *Risk Assessment.* Once the hazards have been identified, the risks they present must be assessed. Risks may involve employee health, safety, or property. The primary inputs for the risk assessment include risk information regarding severity and likelihood of occurrence and criteria for categorizing the resultant risks.

3. *Risk Abatement.* Once the activities hazards have been identified and assessed, risk abatement actions can be pursued. Hazard elimination is always the first choice, followed by engineered safeguards, administrative controls (such as procedures), and finally PPE. When establishing operational controls, consideration is given to the current engineering safeguards, procedures, and PPE already in place, as well as the need for additional controls. Operational controls may focus on elimination of the causes (hazard elimination), or they may emphasize risk reduction assuming the hazard cannot be eliminated.

4. *Monitoring.* Monitoring is initiated, if not already in place, to ensure that the controls implemented are effective in safeguarding the organization's employees. Monitoring may be implemented through periodic operator logging, through event recording, or through periodic safety reviews.

METHODOLOGIES FOR IDENTIFYING JOB HAZARDS

In this section we review some of the more common methods for identifying job hazards. The principal methods include:

Figure 4.5   Area Hazard Identification Checklist (Area)

Hazard Identification Checklist

Area: _____   Date: _____   Analyst: _____

Does the workplace or its surrounding environment present any of the following hazards?

☐ Slips/falls on level surfaces (protrusions, uneven surfaces, holes, wrinkled flooring material, wet or slippery surfaces, oil, cracked concrete or curbs, unguarded pits or shafts)

☐ Falls of persons from heights (lack of guardrails, improper usage of ladders, failure to use harnesses, etc.)

☐ Falls of tools, materials, etc., from heights (lack of kickplates, failure to use restraining devices)

☐ Inadequate tool safeguards (guarding, insulation)

☐ Inadequate headroom

☐ Hazards associated with manual lifting/handling of tools, materials, etc.

☐ Inadequate machine safeguards

☐ Vehicle hazards, e.g., plant vehicles and forklifts

    ☐ No designated walkways for personnel or unsafe layout of forklift travel path

    ☐ Inadequate room/congestion

    ☐ Loading/unloading routes for tractor-trailers

☐ Plant layout (e.g., change rooms in center of plant, congestion)

☐ Fire and explosion hazards (flammable or combustible materials, reactive chemicals, high-voltage electricity)

    ☐ Container safety devices installed (grounding and bonding devices, caps, flash arrestors, etc.)

    ☐ Incompatible materials separated and/or stored separately

    ☐ Fire suppression systems (e.g., sprinkler systems) installed, operational

    ☐ Combustible materials removed

- FMEA
- HAZOP
- JHA

Failure Mode and Effects Analysis

FMEA is a systematic process for analyzing what can go wrong with a product, a process, or a system. Things that could go wrong are called failure modes. The failure modes of concern could be related to product quality issues, equipment, or system reliability; data processing errors; or workplace hazards. Many organizations are

Figure 4.6 Job Hazard Analysis and Risk Assessment

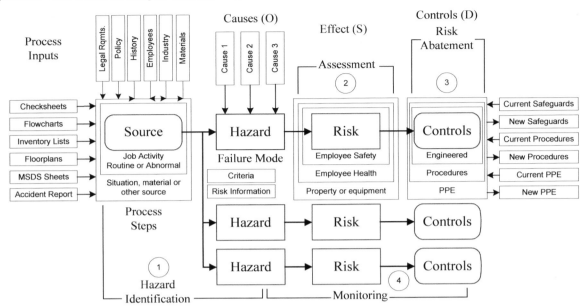

already familiar with the FMEA process and may therefore prefer to use this methodology for hazard analysis and risk assessment. An example of a safety FMEA is shown in figure 4.7.

In the example, each step of the activity (hand grinding) is shown in the first column. The failure modes, or job hazards in this case, are shown in the next column. These hazards would be identified through group discussion or activity observation. The potential safety effects of each hazard are next listed, and each hazard is given a severity ranking on a scale of 1 to 10, with 10 being the most serious. Next the factors that could cause each failure mode are identified and listed. Note that there may be more than one hazard for each step and more than one cause for each hazard. Each cause and hazard combination is then given a likelihood of occurrence ranking based on historical records, industry data, or team experience. This ranking is also based on a scale of 1 to 10, with 10 being a frequent occurrence. Cause prevention controls are then listed. These controls focus on eliminating the cause of the job hazard not elimination of the hazard itself. Note that cause prevention controls are not ranked, because their effectiveness will be evidenced in the occurrence rankings. Failure mode mitigation controls, or controls used to mitigate the effect of the hazard if it does occur, are then listed and an abatement factor assigned based on how effective they would be in protecting the worker.

The risk priority number, or RPN is then calculated by multiplying the severity ranking, occurrence ranking, and abatement ranking for each job hazard, which will result in a value of between 1 and 1,000. This number provides a numerical ranking that can be used to prioritize which jobs require additional operational controls or redesign to eliminate the hazard. Actions to eliminate the hazard, or to lessen its risk if it cannot be eliminated, are then assigned and documented on the FMEA form. The form also has columns for documenting responsibilities for action, due dates, and recording of the final action taken. Finally, the team should reestimate the rankings and RPN after the action has been taken. Ranking assigned to severity, occurrence, and abatement are normally based on tables that provide criteria and examples for each numerical ranking. An example of a severity ranking table is shown in figure 4.8.

The table shown in figure 4.8 is an example. Similar tables would be developed for likelihood of occurrence and abatement effectiveness. In practice the team would develop their own tables and insert their own criteria for ranking. Some teams may use a cost impact, in dollars, for the impact on property and equipment

Figure 4.7 Safety Failure Modes and Effects Analysis

Job Steps/ Activity Details	Potential Failure Mode(s) (Job Hazards)	Potential Effects of Failure (Health & Safety Effects)	Severity	Potential Cause(s)/ Mechanisms Of Failure/Hazard	Occurrence	Current Operational Controls- Cause Prevention	Current Operational Controls- Failure Mode Mitigation	Abatement	RPN	Recommended Actions
Operator picks up casting from box on floor. Each casting weighs 15 pounds, grinds approximately 30 castings per hour.	Drops casting on foot	Fracture of bones in feet	6	Casting slips from hands during lifting	4	None	None	10	240	Provide leather gloves
				Casting strikes edge of table while lifting	3	None	None	10	180	See below
	Strains back while picking up casting	Non-permanent partial disability	5	Box of castings is located on floor slightly behind and to side of operator, castings weigh 15 lbs. each	2	None	Workers are provided with back brace, but use not mandatory.	8	80	Redesign work table and container to eliminate bending and twisting when retrieving casting
	Burr on casting cuts hand	Minor hand laceration	4	Castings have sharp burrs	5	None	None	10	200	Provide leather gloves
	MSD injury from repeated lifting and twisting	Permanent partial disability	7	Operator has to bend and lift approximately thirty 15 pound castings each hour, 8 hours per shift.	3	None	Workers are provided with back brace, but use not mandatory.	8	168	Redesign work table and container to eliminate bending and twisting when retrieving casting
Operator places and grinds castings	Flying particle in eyes	Loss of eyesight	8	Improper positioning of workpiece on grinder	5	Training and work instruction	Chip deflector and Eye protection	3	120	Install limit switch on deflector to lock out if not correctly positioned
				Grinding wheel failure	2	Daily inspection of grinding wheel, proper positioning shown in work instruction	Chip deflector and Eye protection	3	48	
				Improper placement of chip deflector	3	Warning and picture of proper deflector placement	None	10	240	
	Hand or finger contacts grinder	Serious laceration	6	Weight and shape of casting makes it hard to handle	4	None	None	10	240	Provide fixture to support fixture while grinding

instead of the time to repair. Likewise some teams may prefer to use a five-point scale instead of a ten-point scale. The only rule is that once an organization has developed ranking tables and criteria, then they should be used consistently.

The advantages of FMEA methodology are that it is systematic and thorough, and many companies already have experience in the use of the technique. The disadvantage is that it can take more time to complete, requires the use of a team (i.e., more resources), and there can be a temptation to complete the analysis without ever actually observing the job in operation. Although group discussion is a recognized effective method for identifying job hazards, it is dependent on getting the right people involved on the team.

Hazard and Operability Study

A HAZOP identifies hazards and operability problems. The concept involves investigating how a process might deviate from the design intent. If, in the process of identifying problems during a HAZOP, a solution

Figure 4.8   Severity Ranking Table for Safety Failure Modes and Effects Analysis

Effect	Criteria: Severity of Effect Employee Safety or Health	Criteria: Severity of Effect Equipment or Property Damage	Ranking
Catastrophic without Warning	Death or multiple serious casualties, without warning.	Significant property damage leading to loss of facility with no warning.	10
Catastrophic with Warning	Death or multiple serious casualties, with warning.	Significant property damage and loss of facility with warning.	9
Serious	Serious personal injury and permanent and total disability (loss of limbs).	Significant property damage without total loss of the facility.	8
High	Serious personal injury that leads to temporary disability with eventual return to full status (broken bones, severe burns, severe lacerations).	Destruction of or major damage to equipment without impact on surrounding facilities or equipment. Repairs require more than 1 day.	7
Moderate	Significant personal injury that results in lost workdays but no disability (pulls, strains, bruises, cuts, moderate burns, etc.)	Moderate damage to equipment that can be repaired within 1 day.	6
Low	Personal injury that results in the need for first aid, but little appreciable loss of work time (< 2 hours) to address (minor cuts, minor burns, etc.).	Some damage to equipment but it can be easily repaired in less than 1 hour.	5
Very Low	Personal injury that results in the need for first aid, but little appreciable loss of work time (< 30 minutes) to address (minor cuts, minor burns).	No damage to equipment but support required to replace or reset safety feature.	4
Minor	Minor injury that does not require the need for first aid (slight bruise, dust in eye, ect.).	No damage to equipment, but equipment must be restarted or reset by production team member.	3
Very Minor	Nuisance issue—no effect on personal safety or health (slightly elevated humidity or temperature, odor, etc.).	No damage to equipment, but some impact on production capacity other than complete shutdown.	2
None	No discernible effect.	No discernible effect.	1

becomes apparent, it is recorded as part of the HAZOP result. HAZOP is primarily used in process industries, such as chemical manufacturing, but can be adapted for other systems made up of discreet electrical and mechanical components. The author has used it to identify hazards associated with injection molding, chemical mixing, and paint manufacturing. Other applications have included software design and engineering. HAZOP is based on the principle that several experts with different backgrounds can interact and identify more problems when working together than when working separately and combining their results.

The HAZOP concept is to review the plant in a series of meetings, during which a multidisciplinary team methodically brainstorms the process design, following the structure provided by the guide words and the team leader's experience. The team focuses on specific points of the design called study nodes. At each of these study nodes, deviations from expected process parameters are examined using the guide words. The guide words are used to ensure that the design is explored in every conceivable way. Thus the team must identify a fairly large number of deviations, each of which must then be considered so that their potential causes and consequences can be identified.

The best time to conduct a HAZOP is during system design of a process, after concept design but before design finalization. At this point, the design is well enough defined to allow meaningful answers to the questions raised in the HAZOP process. Also, at this point it is still possible to change the design without a major cost. However, HAZOPs can be done at any stage including after installation.

Some HAZOP terms that are important in understanding this methodology include:

- Hazard: Any activity or event that could result in injury to personnel or damage to equipment or facility.
- Operability: Any malfunction or breakdown that would impact regulatory compliance or community relation status or profitability.
- Guide words: Standard terms used to identify deviations from design intent by systematically combining process parameters with guide words. Examples and their meaning include:
  ○ Other Than: Complete substitution of feedstock, part, equipment, or activity, (e.g., Material/Other Than suggesting addition of a completely different material into process);
  ○ Fluctuation: Design intent is achieved only part of the time (e.g., Flow/Fluctuation);
  ○ Early/Late/Before/After: Used when studying sequential operations. Indicates a step is done at the wrong time or out of sequence (e.g., Feed/Early);
  ○ Also: The design intent is achieved but in addition other activity occurs (e.g., Flow/Also indicating contamination in a product stream);
  ○ Reverse: The opposite of the design intent (e.g., Flow/Reverse);
  ○ Other: The activity occurs but not in the way intended (e.g., Maintenance/Other suggesting activity is not done in accordance with the preventive maintenance instructions);
  ○ No/Lack Of: Design intent does not occur or operational aspect is not achieved (e.g., Flow/No, Venting/Lack Of);
  ○ Less/Low: Quantitative decrease in the design intent occurs (e.g., Pressure/Low)
  ○ More/High: Quantitative increase in the design intent occurs (e.g., Temperature/High).
- Process parameters: Attributes of which deviation(s) can have negative effects on safety or environmental compliance and product quality (e.g., material flow, temperature, head pressure, screw/rotor speed, feed location/sequence, mix composition, static/friction, motor current, venting, maintenance, level, time, pH, frequency, viscosity, voltage, separation, reaction, etc).
- Node: A specific portion or section of a process, equipment, activity, as selected for study (e.g., Zone 1 [Twin Screw #2] on an injection molding machine).
- Design intent: Overall purpose of which a process or activity has been designed for (e.g. safe and efficient blending of different materials into a homogeneous mixture, at a controlled speed, rate, or temperature).
- HAZOP Team: Team assembled for conducting study, which includes people knowledgeable about process, equipment, or activity.

The HAZOP process is shown in figure 4.9. The process starts at the upper left-hand corner and cycles until all relevant study nodes have been evaluated.

The study is normally initiated using a system schematic or line block diagram showing all major components and parameters of interest. The analysis would be continued until all other significant sources of hazards were identified and appropriate countermeasures (operational controls) determined. The results would typically be included on a form to serve as a record of the analysis and summary of actions to be taken. An example is provided in figure 4.10.

The advantages of HAZOP, like the FMEA, are its systematic and methodical analysis of what could go

Figure 4.9   Hazard and Operability Study Process

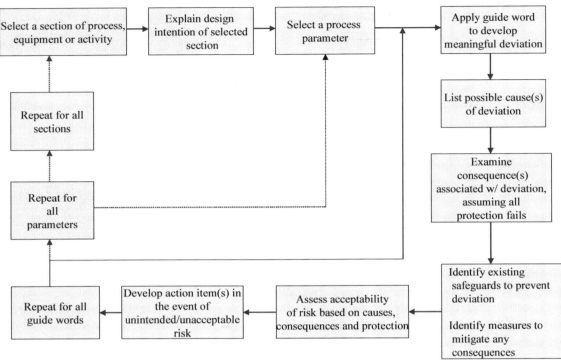

Figure 4.10   Hazard and Operability Study for One Node, for No/Flow

Coordinator: John Doe Scribe: Melissa Manchester Participants: James Band, Dick Hudson, Sally Shield						
Parameter	Guide Word	Causes	Preventive Safeguards	Consequences	Detection Safeguards	Actions/Comments
Flow	No	No feed material in the phosphoric acid storage tank		Excess ammonia in the DAP reactor		Add an alarm/shutdown of the system for low phosphoric acid flow to reactor
		Phosphoric acid feed line valve control valve B fails closed	Periodic maintenance of valve B	Release of unreacted ammonia to the work area and surrounding atmosphere	Ammonia detector & alarm in building	Ensure periodic maintenance and inspection for valve B is adequate
		Plugging of phosphoric acid feed line		Potential injury to employees/local residents due to exposure to ammonia fumes		Provide routine inspection of phosphoric acid feed line
		Rupture of phosphoric acid feed line		Loss of finished DAP product		Enclose DAP storage tank

wrong. It is particularly useful in analyzing systems, especially where there has not been a history of accidents, but where there is a significant potential for injury if an incident occurs resulting from the material or stored energy contained in the system. In instances where the operational and safety history has been relatively good, process operators may have a hard time identifying potential hazards, because there is no experience to draw on. The HAZOP, with its inclusion of process or system design experts and its systematic application of the guide words to each likely point of failure, can help ensure these rare but potentially catastrophic hazards are not overlooked.

The disadvantages of HAZOP are its focus on systems and not tasks and the relatively high amount of effort that will be required to complete the analysis. A HAZOP review of the system used in the example may take a several days for an experienced team. Note that it also does not provide a severity or likelihood of occurrence ranking, rather the assumption is that the accident could occur, and the focus is on preventing the cause or mitigating the effect.

Job Hazard Analysis

JHA is a systematic approach to identifying hazards and assessing their risks. Although there are several variations, the two primary approaches to JHA are group discussion and physical observation.

In the group discussion methodology, a group of functional experts, typically process operators or supervisors, meet and discuss the hazards associated with the task(s) being analyzed. The process experts use their extensive operating experience to identify these hazards and their causes. Various forms of brainstorming are typically used to surface ideas. Actual accidents and near misses can be recalled to further clarify the hazards. The use of a team of experts, rather than one expert alone, results in a reasonable collection of typical hazards found in the workplace. A process flowchart, detailing each step of the process, should be used to help focus the discussion and to ensure all action steps are considered.

The second method involves actual observation of the task being performed. In this case a small team or even a single well-trained individual first observes the task and notes each step required to complete it. The observer(s) then watch the task a second time, noting the hazards that exist during each step. Once all hazards have been identified the observer(s) continue to watch the task, this time focusing on countermeasures (operational controls) that could be used to eliminate the hazards or reduce their likelihood of causing an accident or their effects if one were to occur.

Both methods require some knowledge of typical hazards and controls that can be useful in addressing them. The group discussion method can surface unusual hazards, not typically visible during the time frame of a single observation and which therefore might be missed during physical observation. It can be particularly useful in surfacing hazards associated with nonroutine activities, which are often neglected during the physical observation method because of their infrequent occurrence. The use of a group to perform the analysis builds on the collective wisdom of many operators, including their near misses and observation of the unsafe practices of others. The disadvantage of the group discussion method is that rarely are all of the participants well trained in hazard identification, and hazards identified tend to be those that have evidenced themselves as a result of actual accidents and near misses. Potential, and possibly serious, hazards may not be recognized by the group. Also, the effectiveness of the group discussion method will be seriously affected if the participants are not truly motivated in their desire to initiate changes to make the workplace safer, perhaps because of their good safety record, desire to avoid cumbersome or uncomfortable safety controls, or their belief that safety is primarily the responsibility of each individual worker themself.

The physical observation technique is useful in identifying hazards that may not be recognized during the group discussion method, either because of a lack of training on the part of the participants or because of a lack of history of injuries. Many types of latent injuries or health effects, such as cancers, respiratory issues, and repetitive motion injuries, fall into this category. The physical observers will normally be better trained in the nature and types of hazards and can generally spot them easier than participants in the group technique.

In addition, the systematic and detailed observation of the job ensures that minor but still potentially hazardous steps (such as picking up a fifteen-pound casting) are not overlooked.

The ideal situation would use both the observation and group discussion methods. One or two trained observers could watch the task being performed to identify relevant hazards, and the JHA could then be reviewed by a small group of experts to ensure that nothing was overlooked. The risk assessment (to be discussed later) might also then be performed by this group. The problem with this methodology is the limited availability of resources, because the number of JHAs that will need to be completed will be large in a typical organization. If the organization must select one or the other method, then the physical observation methodology probably works best, because it is easier to spot hazards while actually watching the task being performed. Group discussion could be used to supplement the observation method for tasks that are by their nature high risk.

The steps involved in performing a JHA will now discussed, starting with a description of the methodology as recommended by OSHA. We will then examine variations to the OSHA methodology that have proven to be useful along with a suite of tools that can be used to manage the process. We will assume that the observation method will be used.

The first step is to break down the task into its basic actions. Nearly every job can be broken down into job tasks or steps. When beginning a JHA, watch the employee perform the job and list each step as the worker takes it. Avoid making the breakdown of steps so detailed that it becomes unnecessarily long or so broad that it does not include basic steps. You may find it valuable to get input from other workers who have performed the same job. Later, review the job steps with the employee to make sure that something has not been omitted.

Once the job has been broken down then it is time to identify the hazards associated with each step. JHA is an exercise in detective work; the goal is to discover the following:

- What can go wrong?
- What are the consequences?
- How could it arise?
- What are other contributing factors?
- How likely is it that the hazard will occur?

First identify the steps needed to perform the task. The task is illustrated in figure 4.11.

Step 1. Reach into metal box to right of machine, grasp casting, and carry to wheel.
Step 2. Push casting against wheel to grind off burr.
Step 3. Place finished casting in box to left of machine.

Next identify and document the hazards that could occur during each step. OSHA's format for this analysis is shown in figure 4.12.

Figure 4.11   Grinding by Hand

Figure 4.12a   Job Hazard Analysis for Grinding Operation

Job Location:	Analyst:	Date:
Metal Shop	Joe Safety	2/6/06

Task Description: Worker reaches into metal box to the right of the machine, grasps a 15-pound casting, and carries it to grinding wheel. Worker grinds 20 to 30 castings per hour.

Hazard Description: Picking up a casting, the employee could drop it onto his foot. The casting's weight and height could seriously injure the worker's foot or toes.

Hazard Controls:

1. Remove castings from the box and place them on a table next to the grinder.
2. Wear steel-toe shoes with arch protection.
3. Change protective gloves that allow a better grip.
4. Use a device to pick up castings.

Figure 4.12b   Job Hazard Analysis for Grinding Operation

Job Location:	Analyst:	Date:
Metal Shop	Joe Safety	2/6/06

Task Description: Worker reaches into metal box to the right of the machine, grasps a 15-pound casting, and carries it to grinding wheel. Worker grinds 20 to 30 castings per hour.

Hazard Description: Castings have sharp burrs and edges that can cause severe lacerations.

Hazard Controls:

1. Use a device such as a clamp to pick up castings.
2. Wear cut-resistant gloves that allow a good grip and fit tightly to minimize the chance that they will get caught in grinding wheel.

Figure 4.12c   Job Hazard Analysis for Grinding Operation

Job Location:	Analyst:	Date:
Metal Shop	Joe Safety	2/6/06

Task Description: Worker reaches into metal box to the right of the machine, grasps a 15-pound casting, and carries it to grinding wheel. Worker grinds 20 to 30 castings per hour.

Hazard Description: Reaching, twisting, and lifting 15-pound castings from the floor could result in a muscle strain to the lower back.

Hazard Controls:

1. Move castings from the ground and place them closer to the work zone to minimize lifting. Ideally, place them at waist height or on an adjustable platform or pallet.
2. Train workers not to twist while lifting and reconfigure work stations to minimize twisting during lifts.

The process would be continued until all steps had been analyzed and all hazards identified. The JHA would be complete at this point. The OHSAS 18001 specification requires that further assessment of the risk associated with each hazard be conducted, however, and that brings us to the next step in the process, risk assessment. Note that risk assessment was built into the FMEA process though the assignment of likelihood of occurrence and severity rankings.

When identifying job hazards, an organization should not discount the hazard, because there are current operational controls. Controls fail. People do not always wear their protective equipment. Operators do not always follow procedures. The effectiveness of existing controls will be factored in during the risk assessment. If a hazard is present, document it. Also remember that OHSAS 18001 requires that an organization consider both routine and nonroutine operations, including activities of visitors and contractors.

RISK ASSESSMENT

It is not enough to identify hazards, an organization must also assess the risk that each presents to workers and property. This assessment is important, because an organization will never be able to completely eliminate all risks from the workplace, and there is a cost associated with each operational control. An organization must understand, therefore, which jobs present high enough risks to warrant the expenditure of limited resources in both time and money. The nature of the risks will also drive the number and types of operational controls used to address each hazard. Hazards that could lead to catastrophic injuries or damage should be addressed with multiple layers of controls, including the design of engineered safeguards if their likelihood of occurrence is moderate to high. Other risks, where both the likelihood of occurrence and the severity of the effects are low, may only require simple, inexpensive controls. Additionally, understanding the risks associated with each hazard allows an organization to focus attention on the vital few and not the trivial many.

The most common criteria for assigning risk are likelihood of occurrence and severity of the impact. This was the criteria used by the FMEA process, in this case using a ten-point scale. A more common scale is shown in figure 4.13.

Operational descriptions for each category of severity and likelihood of occurrence and examples of each are then developed. Some typical operational descriptions might include:

- Slightly harmful: superficial injuries, dust in eyes, minor cuts, or nuisance items.
- Harmful: cuts requiring treatment, burns, fractures, sprains, hearing loss, dermatitis, or concussions.
- Extremely harmful: loss of limb, major fractures, disabling injuries, fatalities, cancer, or poisonings.

Figure 4.13   Risk Criteria Table

Likelihood of Occurrence	Severity of Hazard		
	Slightly Harmful	Harmful	Extremely Harmful
Extremely Unlikely	Negligible Risk (N)	Tolerable Risk (T)	Moderate Risk (M)
Unlikely	Tolerable Risk (T)	Moderate Risk (M)	High Risk (H)
Likely	Moderate Risk (M)	High Risk (H)	Unacceptable Risk (U)

- Extremely unlikely: Never happened before, multiple levels of control, or extremely low industry incidence level.
- Unlikely: Rare occurrence in past or low level of industry incidents, only one or two safeguards in place.
- Likely: History of occurrence, high industry rate, or no safeguards in place or safeguards with a history of tampering.

To make it easier to rank the resultant risks, a numerical scale is often assigned to each category and each cell in the table. I prefer to use a 1–3–9 scale, because this provides better separation between the vital few and the trivial many. The multiplicative sum of the likelihood of occurrence and severity rankings will then provide a numerical score for each hazard assessed as shown in figure 4.14.

The result of the risk assessment will be a numerical ranking that associates each job hazard with one of several risk categories, in this case negligible risk, tolerable risk, moderate risk, high risk, and unacceptable risk. These categories then form the basis for action. Note that some variations do not use the numerical rankings, whereas others only use four categories instead of five. These are minor details and the organization should use the methods most comfortable for them. The important thing is to have some rationale, systematic method to assess the risks and to categorize them so that they can be addressed in a reasonable fashion. This is required by the OHSAS 18001 specification.

To conduct the risk assessment, the team considers each hazard identified during the JHA and assigns a likelihood of occurrence ranking based on the operational descriptions for each category. Information that can be used to make the assignments can include:

1. the organization's accident history, including near misses,
2. industry data, if available,
3. OSHA statistics,
4. insurance statistics, and
5. team experience (subjective).

In most cases the team will use a combination of 1 and 5 owing to the absence of specific data for the hazard being considered. This is fine as long as a reasonable estimate is made using a team of operators familiar with the process.

Figure 4.14 Risk Numerical Rankings

Severity of Hazard

Likelihood of Occurrence		Slightly Harmful 1	Harmful 3	Extremely Harmful 9
Extremely Unlikely	1	Negligible Risk (N) 1	Tolerable Risk (T) 3	Moderate Risk (M) 9
Unlikely	3	Tolerable Risk (T) 3	Moderate Risk (M) 9	High Risk (H) 27
Likely	9	Moderate Risk (M) 9	High Risk (H) 27	Unacceptable Risk (U) 81

Severity ranking will then be assigned by the team using the same sources, supplemented by safety engineering expertise if available. It is best to think of the most likely result of the incident, not the worst thing that could possibly happen given the right conditions. In other words, consider first-order effects. Do not take this kind of logical path: *"The incident could result in a minor cut, which if not properly cleaned could result in infection, which if not given proper medical treatment could result in heart disease and death for workers with poor immunity systems."* Anything could probably be shaped into a deadly event, given sufficient creativity. An organization should be reasonable in classification.

Now that we have explored methods used to conduct JHA and risk assessment we can lay out a process for conducting these analyses. At this point the team to be used has been assigned, tools developed, and team trained on the methods that will be used. Before the reviews can begin, we must first identify the tasks that need to be analyzed and prioritize their completion.

Creating the Job Lists

Job lists will need to be created for each department or area within the company. These lists will serve to identify the different types of tasks performed within the area, both routine and nonroutine, along with the priority for analysis. The lists can also be used to track completion of JHA and risk assessments and can help identify objectives for JHA completion. An example of a typical job list is shown in figure 4.15 and is included in the OH&S planning workbook provided on the CD.

In this format jobs are broken down as routine and nonroutine. Each task is assigned a job number, the equipment and materials required to perform the task, and the function(s) involved in performing the activity are indicated. The remainder of the information on the spreadsheet is used to prioritize the tasks and track completion of the JHAs. Notes are embedded in the header cells to explain the column and to help assign priorities. An example is shown in figure 4.16.

One spreadsheet (or other type of list) should be developed by each department or functional manager. The management appointee or project leader will quickly find out how closely the management team has embraced the project at this point. Although the development of the job list is not difficult and can probably be accomplished in a few hours, inevitably some managers will procrastinate on this task. It may be necessary to have one of the CFT team members schedule time to sit down with the area or department manager (or supervisor) to complete it together. The job lists are critical; an organization cannot manage your JHA program without it. Neither can one individual do a credible job of creating job lists for other departments or areas where they are not assigned. Demand accountability for the completion of this task.

The pop-up note in figure 4.16 shows the criteria used to prioritize the jobs on this departments job list. Once the lists are complete, the management appointee and CFT, in consultation with the management team should set goals for JHA completion. A reasonable goal might be to complete initial analysis of all priority one jobs within the next four months. In our example this would require the completion of eleven JHAs or an average of about three a month. This is certainly a reasonable target.

COMPLETING THE JOB HAZARD ANALYSIS AND RISK ASSESSMENTS: USE OF THE OCCUPATIONAL HEALTH AND SAFETY PLANNING WORKBOOK

With a trained team composed of representatives from each area, a prioritized job list of tasks to be analyzed, and goals for the completion of the JHAs, the analysis can begin. Because the methodologies for conducting the JHAs have already been discussed there is no need to repeat it here. Instead we will focus on the use of the tools developed to assist in the completion and management of the assessments.

Figure 4.15   Typical Job List: Warehouse

Area: Warehouse

No.	Job or Activity	Equipment Used	Hazardous Materials Involved	Functions Involved	Job-Specific Safety Procedure or Program?	Pri. 1/2/3	Analyzed?	Status	Current Risk Category	Action Due Date
								R-Y-G	N-T-M-H	
	Routine									
101	Loading/Unloading trailers w/ Forklift	Forklift		Material Handler	Yes	2	No	Y		
102	Manual loading/unloading trailers	Portable conveyer		Material Handler		1	Yes	G	T	
103	Packing and labeling cartons for shipment	Tape dispenser, label gun		Warehouseman		3	No	Y		
104	Unpacking cartons	Box knife		Warehouseman		2	No	R		
105	Applying shrink wrap to pallets	Shrink wrap machine		Material Handler		1	Yes	G	T	
106	Assembly of product racking	Pnuematic hand tools		Warehouseman		1	Yes	G	M	11/5/2006
107	Tear down of fixed racking	Pnuematic hand tools		Warehouseman		1	Yes	G	M	11/7/2006
108	Safe operation of Order Picker	Order Picker		Material Handler	Yes	2	No	Y		
109	Safe operation of Forklift	Forklift		Material Handler	Yes	2	No	Y		
110	Floor cleaning with broom	Broom and mop		Warehouseman		3	No	Y		
111	Floor cleaning with Power Washer	Power Washer	Cleaning agent	Warehouseman		2	Yes	G	T	
112	Manual picking of products	Pick cart, barcode reader		Warehouseman		2	Yes	G	H	11/3/2006
113	Use of computer terminal	Computer terminal		Warehouseman		3	No	Y		
114	Stacking and disposal of wooden pallets	Forklift		Material Handler		2	Yes	G	T	
115	Crating	Hand tools, table saw		Warehouseman		1	Yes	G	T	
116	Operation of the waste compactor	Waste compactor		Material Handler, Warehouseman		1	Yes	G	M	11/23/2006
119	Proper use of table saw	Table saw		Warehouseman	Yes	2	Yes	G	T	
	Non-routine									
150	Spill cleanup	Spill kit		Material Handler, Warehouseman	Yes	2	No	Y		
151	Proper use of eye-wash station	Eye-wash station		Material Handler, Warehouseman	Yes	2	No	Y		
152	Proper use of fire extinguisher	Fire extinguisher		Material Handler, Warehouseman		1	Yes	G	H	11/6/2006
153	Emergency shutoff of storm drains	Storm drain cuttoff valves		Material Handler, Warehouseman		2	No	R		
154	Replacement of Forklift propane tanks	Forklift, handtools	Propane	Material Handler		1	Yes	G	H	11/8/2006
155	Charging of Order Picker batteries	Order Picker, handtools	Electric batteries	Material Handler		1	Yes	G	N	

Performance Monitoring and Reporting				
% Red	9.5%	No. Red	2	
% Yellow	33.3%	No. Yellow	7	
% Green	57.1%	No. Green	12	
		Total	21	

Note that the range in each calculation must be revised whenever new tasks/activities are added or deleted to ensure the calculations are correct.

Figure 4.16   Example of Spreadsheet Embedded Notes

Job-Specific Safety Procedure or Program?	Pri. 1/2/3	A
Yes	2	
	1	
	3	
	2	
	1	
	1	
	1	
Yes	2	
Yes	2	
	3	
	2	
	2	
	3	No	Y

Priority for Evaluation

Priority 1 - Jobs or activities with a history of accidents, near-misses, or lost workdays, jobs involving exposure to dangerous chemicals or voltages/current, jobs requiring working from heights above 3 feet, jobs involving mechanized equipment (other than small hand tools) with the capability for serious injury, or any other activity considered to be inherently dangerous.

Priority 2 - Jobs or activities with no history of accidents, near-misses, or lost workdays, but considered to be potential sources of non-serious injury or harm. Also Priority 1 activities where there are activity-specific safety instructions (e.g. lockout-tagout) in use and under a periodic surveillance program.

Priority 3 - Jobs or activities with no history of accidents, near-misses, or lost workdays, and considered to have little potential for serious injury or harm.

The primary tool provided on the CD is the OH&S planning workbook. This is a Microsoft Excel workbook containing a series of templates designed to help an organization manage its health and safety management program. Some of these tools have already been discussed; others will be discussed in this section; and the remainder will be reviewed in subsequent chapters. At this point we will review those that can be used to assist in the JHA and risks assessments. It is recommended that you open the file named OH&S planning workbook to follow along during this discussion.

The workbook is made up of a series of worksheets, each shown by a tab at the bottom of the workbook. In this section we will discuss the use of the first three worksheets:

- Job List worksheet,
- Job Hazard Analysis and Risk Assessment worksheet, and
- Job Hazard form.

Job List Worksheet

The job list was discussed in the previous section. A separate job list should be prepared for each department or functional area. Blank copies of the job list on the CD can be made using Excel's Edit/Move or Copy/Create a Copy function from the main menu.

Job Hazard Analysis and Risk Assessment Worksheet

The JHA and risk assessment worksheet is an optional tool for documenting and tracking the results of the JHA and risk assessments. The OSHA example previously presented has been shown on the worksheet for purposes of illustration. The worksheet is shown in figure 4.17.

The top row provides the name of the task that was analyzed. The next row provides the name of the function associated with the task. Column 3 shows the steps analyzed in the process that had recognizable hazards. To improve the presentation of the data, the top two rows can be merged once all of the hazards have been identified.

The hazards associated with each step are noted in the fourth row. The table inserted in the far left column lists the generic types of hazards using common codes. This table can provide guidance to the group assigned to conduct the JHA when the group method is used.

The next five rows provide detail, in the form of simple yes/no answers, about the existing controls in use at the workstation. This information will help the team assign risk rankings, which is the subject of the next four rows. Pop-up notes are embedded within the rightmost column to provide the operational definitions when the worksheet is used to facilitate the risk assessment. The user places their mouse over the cell to activate the notes. An example is shown in figure 4.18.

The use of the remainder of the worksheet deals with actions taken to reduce the risks associated with the identified hazards. These sections will be discussed in subsequent sections. Finally note that there is a built-in macro to record the numbers and percentage of the various categories of hazard categories. This can be a valuable method to gauge performance in improving an organization's health and safety posture.

The worksheet is a useful tool for documenting and managing an organization's JHA and risk assessments. However, it can be cumbersome to use if the physical observation method for performing the JHA is used, because it would necessitate transcription of the information on the JHA form (described next) to the spreadsheet. When the group discussion method is used, however, the worksheet provides both the structure and the criteria needed to facilitate and document the analysis. In this situation, a projector should be used during the group analysis and the worksheet projected onto a wall or screen so that the pop-up notes, hazard categories, and other information can be seen by all team members. The sheet would be filled in by the facilitator during the group meeting as hazards are identified.

Figure 4.17   Job Hazard Analysis and Risk Assessment Worksheet

	Hand Grinding					
Primary Activity or Source (Note - always leave one blank column at end of last step)						
Functions Involved	Machinist					
Steps with Identifiable Hazards	Pick up casting	Grasp and hold casting while lifting	Pick up castings			
Hazard associated with step SB – Struck By SA – Struck Against CW - Contact With Cby- Contacted By CI – Caught In CO – Caught On CB – Caught Between O – Overexertion Rep. Motion E - Exposure FS – Fall Same Level FB – Fall to Below	SB - Worker drops 15 lb. casting on foot	CW - Worker contacts sharp burrs causing lacerations	RM - Worker twists or strains while lifting 15 lb. casting			
Step Specifically **Regulated** (Y/N)?	N	Y	N			
Engineered **Safeguards** (Y/N)?	N	N	N			
Procedure Controls (Y/N)?	N	N	N			
Trained for Safe Performance Y/N)?	N	N	N			
Personal Protective Equip. (Y/N)?	N	N	N			
Severity of Hazard (1/3/9)	3	3	3			
Likelihood of **Occurrence**, considering existing controls and history (1/3/9)	3	9	9			
Total Risk **Ranking** (SxL)	9	27	27	0	0	0
Initial Category Based on Ranking (I/H/M/T/N)	M	H	H			
Action (I/M/E) - Reference Job Actions Sheet	E	I	I			
Engineered **Safeguards** (Y/N)?	Y	N	Y			
Procedure Controls (Y/N)?	N	N	N			
Trained for Safe Performance Y/N)?	Y	Y	Y			
Personal Protective Equip. (Y/N)?	Y	Y	N			
Recalculated Severity after action (1/3/9)	3	3	1			
Recalculated Occurence after action (1/3/9)	1	1	1			
Recalculated Total Risk **Ranking** (SxL)	3	3	1	0	0	0
Current Category Based on Ranking (I/H/M/T/N)	T	T	N			
Summary of Resultant Hazard Controls - List specific procedures (by number), PPE and safeguards in place to prevent worker injury or illness or workplace damage.	Redesigned work table. Provided gloves with better grip.	Provided cut resistant gloves.	Redesigned work table to require minimal lifting.			

Monitoring and Reporting Statistics			
% N	33.3	No. N	1
% T	66.7	No. T	2
% M	0.0	No. M	0
% H	0.0	No. H	0
		Total	3

Figure 4.18   Example of Pop-up Note Describing Risk Ranking Criteria

7	Procedure Controls (Y/N)?	N	N	N			
8	Trained for Safe Performance Y/N)?	N	N	N			
9	Personal Protective Equip. (Y/N)?						
10	Severity of Hazard (1/3/9)						
11	Likelihood of Occurrence, considering existing controls and history (1/3/9)						
12	Total Risk Ranking (SxL)						
13	Initial Category Based on Ranking (I/H/M/T/N)						
14	Action (I/M/E) - Reference Job Actions Sheet						
15	Engineered Safeguards (Y/N)?						
16	Procedure Controls (Y/N)?						
17	Trained for Safe Performance Y/N)?						
18	Personal Protective Equip. (Y/N)?	Y	Y	N			
19	Recalculated Severity after action (1/3/9)	3	3	1			

> **Severity Rankings:**
> 9 - **Extremely harmful:** amputations, major fractures, poisonings, multiple injuries, fatalities, cancer, other life-threatening diseases.
> 3 - **Harmful:** lacerations, burns, concussion, serious sprains, non-major fractures, deafness, dermatitis, long-term repetitive motion types of injuries.
> 1 - **Slightly Harmful:** superficial injuries, minor cuts and bruises, eye irratation from dust, nusance and irritation, short-term discomfort.

One challenge with using the worksheet is that it can become extremely large in cases where there are twenty or thirty tasks to be analyzed, each with five to ten steps (or more). To manage this situation, it is recommended that each department have its own workbook containing its job list and JHA and risk assessment worksheets. For a small department with only a few tasks, all of the JHAs could be recorded on the same worksheet. For a larger area or department (such as production) a separate worksheet could be created for each primary activity, such as machining, welding, and assembly. If a worksheet unexpectedly gets too large, simply divide the processes into smaller units.

Job Hazard Analysis Form

The JHA form is an alternative to the use of the JHA and risk assessment worksheet. It would be used by teams using physical observation to conduct the JHA. Discussed briefly in a previous section, the JHA form is reproduced in figure 4.19, this time completed.

This form can be completed at the work site. Note that the risk assessment has been built into the form, as have the actions proposed to reduce the risks. Hazard categories and criteria for ranking the severity and likelihood of occurrence are shown along the bottom of the form.

Although this form does not provide for a centralized file of JHA and risk assessments as does the previous worksheet, it does provide an advantage over that worksheet in that it can be posted directly at the work site when in its final form. The completed JHA can then be used to train new operators and help maintain an awareness of hazards and related controls unique to the job.

Which format should be used? It is really up to the organization as a matter of preference. I recommend the use of the JHA and risk assessment worksheet when the group discussion method is used to identify job hazards and the JHA form when the direct observation method is used. As an alternate, the team may elect to use the JHA form for conducting the on-site hazard analysis with transfer of the information onto the JHA and risk assessment worksheet afterward. This would provide an organization the benefits of better management and tracking of JHA completion with the worksheets and the ability to communicate the risk analysis to employees through posting of the JHA form at the work site. The cost, of course, is the additional effort required to transcribe the information from the JHA form onto the worksheet.

Most of the organizations that I have worked with have preferred to use the JHA form over the spreadsheet. They like the fact that it can be used to perform both JHA and risk assessment on the same form, that it can be taken to the work site and used to conduct the JHA, and that the finalized form can be posted at the

Figure 4.19 Completed Job Hazard Analysis Form

Plant: Breking Road		Department: Machining			
Job/Task: Hand Grinding	Page: 1 of 1		Updated:	Initial ☑ / Final ☐ / Revision ☐	
Position(s) Performing The Job/Task: Machinist	Supervisor(s): Bentley		JSA Performed by: O"Conner, Tomlins		
Required Equipment for the job:	Stationary Grinding Machine				
Job specific required P.P.E.:	Safety Glasses				
Job specific optional P.P.E.:	None Specified				

	Job Procedures	Potential Hazards	Severity (1-3-9)	Likelihood (1-3-9)	Total (1-81)	Category	Action/Procedure to Control/Eliminate
1	Pick up casting	Worker drops 15 pound casting on foot	3	3	9	M	Reposition bracket than can interfere with handling. Provide gloves with slip resistance finish. Require steel toed shoes.
2		Worker twists or strains back while lifting 15 pound casting	3	9	27	H	Consider raising box with castings off the floor to eliminate bending. Provide training in lifting. Provide back support.
3	Grasp and hold casting while grinding	Worker contacts sharp burrs causing lacerations	3	9	27	H	Provide operator with protective gloves. Note recommendation 1.
4		Workers hand comes into contact with grinding wheel	3	9	27	H	Provide operator with protective gloves. Note recommendation 1. Better position wheel guard - too much clearance.
5		Grinding fragments puncture workers eyes	3	3	9	M	Currently require safety goggles and have a grinding guard that deflects some chips. Consider better positioning of guard.
6	Place casting in bin	Same as items 1 and 2					Same as actions for item 1.
	Overall Rating (highest individual ranking):					HIGH	

Potential Hazards: SB-Struck By, CW-Contact With, Cby-Contact By, CB-Caught Between, SA-Struck Against, CI-Caught In, CO-Caught On, O-Overexertion or Repetitive Motion, FS-Fall at Same Level, FA-Fall From Above, E-Exposure to Chemicals, Vapor or Noise

Severity Rankings: 1 (Slightly Harmful) - superficial injuries, dust in eyes, minor cuts. 3 (Harmful) - cuts requiring treatment, burns, fractures, sprains, dermatistis, concusions. 9 (Extremely Harmful) - loss of limb, major fractures, disabling inju

Liklihood Rankings: 1 (Extremely Unlikely) - Never happened before, muliple levels of control, extremely low industry occurences. 3 (Unlikely) - Rare occurrence in past or near misses, at least one safeguard in place. 9 (Likely) - History of occuren

Risk / Hazard Category: 1- Neglible (N) 3- Tolerable (T) 9- Moderate (M) 27- High (H) 81- Unacceptable (U - Do not operate until risk is reduced):

work site as a means of increasing worker awareness of the hazards and corresponding controls associated with the tasks they perform.

Identifying Operational Controls

The next step after identifying and assessing workplace hazards is to assign operational controls and metrics. In many cases adequate controls will already exist. In others, controls will need to be improved or developed to provide safeguards against worker injury. What is needed is some method to help determine where actions should be taken.

Figure 4.20 shows an example of criteria for action that can be used to determine and prioritize the expenditure of resources on risk reduction. This table is shown next to the risk categorization criteria presented previously to demonstrate how the two work together. The five categories were determined based on the risk assessment scores calculated for each recognizable hazard. The table on the right then provides guidelines for action based on these categories. An explanation for the actions for each category is as follows:

- *Unacceptable risk*: This is a combination of an extremely harmful severity that is likely to happen. Such risks cannot be accepted. The action for this rank is immediately stop the process and implement operational controls to reduce the risk before allowing operations to continue.

Figure 4.20   Criteria for Determining Response to Hazard Categories

Risk Category	Action
Negligible Risk	No action or documentation required
Tolerable Risk	No additional action required, monitor to ensure risk remains tolerable
Moderate Risk	Reduce risks through appropriate improvements as resources permit.
High Risk	Reduce risks immediately.
Unacceptable Risk	Do not allow work to proceed until risk has been reduced to moderate levels.

Severity of Hazard

Likelihood of Occurrence		Slightly Harmful 1	Harmful 3	Extremely Harmful 9
Extremely Unlikely	1	Negligible Risk (N) 1	Tolerable Risk (T) 3	Moderate Risk (M) 9
Unlikely	3	Tolerable Risk (T) 3	Moderate Risk (M) 9	High Risk (H) 27
Likely	9	Moderate Risk (M) 9	High Risk (H) 27	Unacceptable Risk (U) 81

- *High risk*: The result of an extremely harmful consequence that is unlikely (but not remote) or a harmful consequence that is likely. Take actions to reduce risks immediately. The process does not have to be shut down, but quick action is needed to ensure worker safety. This category is typically the highest priority for action.
- *Moderate risk*: There are three combinations of severity versus likelihood that can fall into this category. Because it is not high risk, immediate action is not needed, but improvements should be considered as resources permit. Many of the organization's risks will probably fall into this category. Improvements should be made but not before attacking high-risk activities.
- *Tolerable risk*: Activities in this category may be harmful, but they are extremely unlikely, or they are only slightly harmful and unlikely. Normally these risks can be lived with, and no additional action is required assuming they stay at this level. If the organization were successful in reducing all of their high- and moderate-risk activities to this level, then improvements would begin to focus on risk reduction for these activities.
- *Negligible risk*: These are only slightly harmful and extremely unlikely. No action or documentation, other than the initial JHA and risk analysis that showed them to be negligible, is needed.

Notice that the criteria shown in figure 4.20 were called guidelines. The reason for this is that invariably situations will arise where tolerable or moderate-risk activities will be identified along with inexpensive, readily available means to reduce their risks. As an example consider the case of providing protective gloves to the machinist at the grinding (by hand) station. Because there will probably be some high-risk activities still awaiting operational controls, does this mean that an organization cannot, or should not, provide the gloves to the machinist? Of course not. *If risk reduction is easily accomplished then it should be taken no matter what the category.* The criteria serve to help guide decision makers when the necessary controls are neither inexpensive nor easy to implement. Because all companies are faced with limited resources, some means of prioritizing the distribution of those resources is required.

With this foundation we are now ready to discuss the different types of operational controls and the hierarchy governing their selection.

HAZARD ELIMINATION

This first type of control is not really a control at all, rather it seeks to eliminate the hazard completely. Hazard elimination is always the first option that should be considered in accordance with the philosophy: *Where no hazard exists there can be no accident.* While it will not be possible to eliminate the hazard in many if not most cases, it should always be considered first before proceeding to the other options. Examples of hazard elimination include:

- Using pneumatic control instead of electrical control in environments where combustible fumes or fire dangers exist.
- Replacing combustible materials with noncombustible materials where possible.
- Replacing hazardous materials with nonhazardous alternatives where possible.
- Replacing hoses with many fittings that can leak and cause slips, falls, or even fires with continuous lines with minimal connectors.
- Rounding or padding sharp corners or edges on desks and equipment.

Focused efforts to eliminate hazards rather than simply reducing their risks are a signal that the organization has a mature and proactive health and safety program.

ENGINEERED CONTROLS

Where hazards cannot be eliminated next consider engineered controls. These include:

- Automatic safety devices, such as pressure relief valves, interlocks, blowout plugs, sensor-operated air handling systems, fuses and breakers, and a wide variety of sensor circuits.
- Passive safety features, such as thermal insulation, explosion-proof containers, electrical equipment (for use in flammable atmospheres), level sensors, and permanently installed nonconductive matting.
- Isolation including cages, physical barriers, deflectors, containment buildings, relocation of equipment or operations to less populated areas, installation above normal working level, and lockouts.
- Alarm and detection circuits, continuous monitors (gas monitors, level indicator and alarm circuits, infrared detectors, and motion sensors).

Engineered controls are the most preferred method of risk reduction, because they do not rely on human action to function (other than proper installation and maintenance).

ADMINISTRATIVE CONTROLS

Where hazards cannot be eliminated or addressed with engineered controls, then administrative controls should be considered. These controls rely on human intervention or compliance to be effective. Examples include:

- Procedures, policies, and instructions.
- Programs, such as lockout-tagout and permit required confined space entry.
- Activity specific training, such as confined space entry training.
- Buddy systems, where two operators are assigned to potentially hazardous activities, such as confined space entry and welding fire watch. The second operator stays outside the danger zone and stands by to notify others in case of emergency or to take specific authorized immediate action.
- Warning labels and postings.
- Job rotation (to prevent fatigue or disinterest).
- Requirements to check a circuit de-energized and to discharge all capacitive circuits prior to commencing work on recently de-energized electrical equipment.

PERSONAL PROTECTIVE EQUIPMENT

The use of PPE is the last option and should be used to supplement the preceding options or in situations where the other controls are not possible. The use of PPE as a supplement to other controls may include:

- The use of respirators to provide additional protection to workers in atmospheres that may be contaminated with harmful fumes, such as paint booths.
- The use of leather gloves to protect the machinist's hands during grinding.
- The use of apron, face shields, and chemical-resistant gloves to protect a chemist during mixing operations.
- The use of hard hats in a construction zone.
- The wearing of eye protection in a manufacturing environment as a secondary form of protection against flying chips and debris.
- The use of nonconductive gloves during maintenance on a piece of electrical equipment.

The use of PPE in these instances does not replace other forms of operational controls but rather adds an additional layer of protection to the worker. Because the wearing of PPE can make workers uncomfortable, many may not wear it. A robust safety culture is required to promote the proper wearing of PPE.

In some situations PPE may be the primary form of protection, but these are, with some exceptions, restricted to emergency response situations. An example is first responders' reliance on oxygen-air-breathing apparatus.

Note that the use of PPE requires in all cases some level of administrative controls to ensure it will serve to protect the employee using it. Examples of these administrative controls include:

- Proper selection and acquisition of PPE that is rated for the environment in which it will be used. There are many different types of respirators, hard hats, gloves, and other PPE, each rated for a specific application. Controls must be in place to ensure the PPE selected will be appropriate for the hazard it may have to protect against.
- Training on the proper use of PPE. PPE is often ineffective, because the wearer does not know how to fit or adjust the PPE. This is particularly true of respirators and emergency-air-breathing apparatus, but it can also be true for anticontamination clothing and even ear plugs.
- Policies and procedures on the proper inspection, maintenance, cleaning, and storage of PPE.

Once the hazards have been identified and their risks assessed, the team should focus its attention on the need for any additional operational controls required to lessen the risk, using common sense and the guidelines provided in figure 4.20. Figure 4.21 summarizes the selection hierarchy for identifying operational controls.

The team should record the controls that have been used to reduce the risk and recalculate the new risk category considering these new controls. The results should be shown on the JHA and risk assessment worksheet or on the final JHA form. The JHA form can then be posted, if used, at the work site to ensure all operators are aware of the hazards associated with the task and the methods used to protect against injury.

Summary

In this chapter we learned how to plan and conduct JHA and assess the resultant risk. The chapter started with the identification and listing of the legal requirements that the organization must comply with in accordance with federal, state, or local regulations and standards. JHA was then discussed beginning with the selec-

Figure 4.21 Hierarchy for Selection of Operational Controls

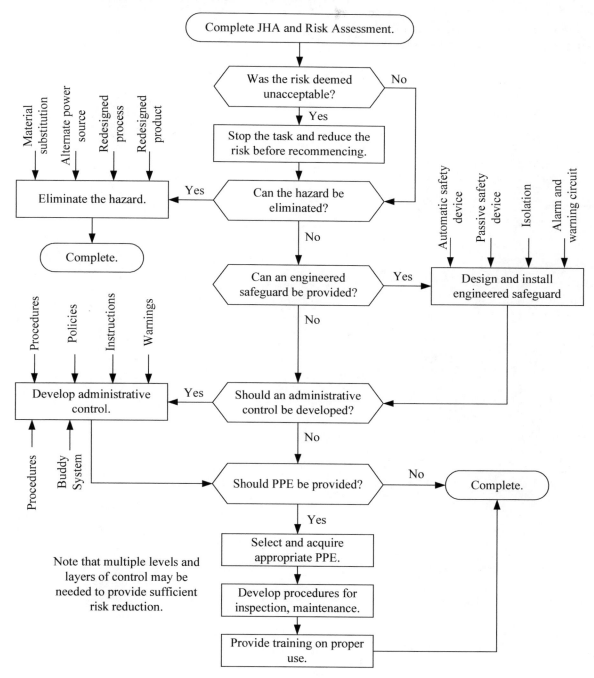

tion and training of team members who will perform the analysis and the development of tools and forms that they will use. The creation of departmental or area job lists and the prioritization of jobs to be analyzed was also discussed. Several different methodologies were presented for conducting the JHA, and criteria were provided for completing the subsequent risk assessment. Finally, the selection of operational controls was reviewed for those areas where the risk assessment indicates a need to reduce the risk to employees.

CHAPTER 5

Developing Key Support Processes

This chapter will focus on the design and development of the common supporting processes needed to maintain and improve the HSMS. Those with existing quality or environmental management systems already in place will find that many of these processes have already been developed and can be used either as a template for their OH&S procedures or as a place to put the expanded controls needed to manage the HSMS common elements. Wherever possible I recommend developing integrated processes for common elements, because it normally results in less confusion, better systems, and less effort to maintain compared to separate processes for the quality, environmental, or occupational HSMSs.

Document and Record Control Process

Chapter 2 presented the requirements relating to document and record control. It was noted that the OHSAS 18001 specification requires only minimal documentation, and a discussion was presented to help guide the organization in its determination of where and when written procedures should be developed. It is important that documented OH&S procedures, where developed, be adequately controlled. Normally the first step in establishing control is to develop a master listing of the procedures, instructions, forms, and other documents that form the basis for the management system. This is normally called the master list. An example of a simple master list is shown in figure 5.1.

The example shows the different categories of documents and the information on how each is controlled. The numbering system shown is simple and can be used with systems that do not have too many procedures. In this format HSP stands for OH&S system procedure, HSI for OH&S system instructions, and HSF for OH&S system form. Documents are numbered sequentially starting with 001. External documents use a control number as assigned by the agency that issued them (e.g., the Michigan DEQ form number in this example). Many other formats are possible, but the important thing is for each document to have a unique identifier, typically by both name and number but sometimes by name only, to ensure documents are not confused.

The second column includes the current revision date of the document. This is probably the most important piece of information on the master list, because it can be used to ensure the most current (and therefore correct) revision is being used. In addition the list shows where the documents are located, both in electronic and hard-copy forms. This allows for updating of documents and removal of obsolete versions when documents are revised. Retention periods are also shown, although these generally apply more to records than documents. Finally, any documents that are retained after they are obsolete are shown, and all records have their storage locations identified where they can be retrieved if needed.

The example shows only a few of the documents and records that would be needed for a typical HSMS. For larger companies or those with more complex operations, it is often best to create a separate workbook or database to use as the master list. The example also shows the typical structure of most formal document

Figure 5.1   Simple Master List of Documents and Records

Document Name	Number	Current Revision	Locations	Retention Period	Record Storage
Procedures					
Health & Safety Policy Manual	HSP-001	4/16/2006	P//HSMS/Procedures	Superceeded	NA
Job Hazard Analysis and Risk Assessment Procedure	HSP-002	5/2/2006	P//HSMS/Procedures	Until Superceeded	NA
Internal Auditing	HSP-003	5/10/2006	P//HSMS/Procedures	Superceeded	NA
Document and Record Control	HSP-004	5/10/2006	P//HSMS/Procedures	Superceeded	NA
Material and Waste Handling	HSP-005	6/14/2006	P//HSMS/Procedures Shipping office Maintenance	Until Superceeded	NA
Job Safety Handbook	HSP-006	11/5/2005	P//HSMS/Procedures Each Employee	7 years	P//HSMS/Obsolete
Instructions					
Lift Truck Battery Charging	HSI-001	12/15/2005	P//HSMS//Instructions Facilities Supervisor	Until Superceeded	NA
Bulk Liquid Unloading	HSI-002	7/1/2006	P//HSMS//Instructions Maintenance	Until Superceeded	NA
Lockout-Tagout 150 Ton Press	HSI-003	7/21/2004	P//HSMS//Instructions Maintenance	Until Superceeded	NA
External Documents					
150 Ton Press Technical Manual	NA	9/3/2004	Maintenance	Superceeded	NA
City of Oakly Fire Safety Code	NA	5/23/2005	Safety Coordinator	Superceeded	NA
Respirator Training Guide	NA	1/6/2005	Safety Coordinator	Superceeded	NA
Forms and Records					
Communication Log	HSF-001	5/19/2006	Management Appointee	3 years	Mgmt App. Office
Management Review Report	HSF-002	5/19/2006	Management Appointee	3 years	Mgmt App. Office
Weekly Safety Committee Report Form	HSF-003	5/10/2006	Safety Coordinator	1 year	Safety Coordinator
MIOSHA 300 Log	PA-00341-A	NA	Safety Coordinator	3 years	Safety Coordinator

management systems, with a level-one policy manual, level-two procedures, level-three instructions, and level-four forms, records, and other documentation. This structure is shown in figure 5.2.

The policy manual serves primarily as a roadmap, outlining the basic structure of the HSMS and guiding the reader to other specific documents for more details on how, who, or when. An example of an OHSAS 18001 policy manual is provided on the CD. Keep in mind that the exact format of the policy manual, or even whether it is one document or the compilation of many, is up to an organization. As a minimum it must provide a general description of the elements that make up the HSMS, how these elements interact, and it must reference other related documents. It also serves as a good place to put information on certain activities that need to be described but which do not warrant their own stand-alone procedure or instruction. The timing, attendance, and reporting of the management review is one such example.

Procedures (also called level-two documents) generally describe core processes that span multiple departments or that are made up of several distinct activities. They provide a general description of what, details on who and when, and generally serve to integrate the various departments or activities together. Not every clause in the specification requires a separate procedure. As previously noted, some do not require a documented procedure at all. I recommend combining clause activities together when it makes sense. For example, I often address clause 4.3.1, 4.3.2, 4.3.3, and 4.3.4 in a single planning procedure, because in many companies these activities are related. Likewise, document and record control can usually be combined, as can internal audits and evaluation of compliance. Examples of common procedures are provided on the CD.

Figure 5.2   Typical Document Management Structure

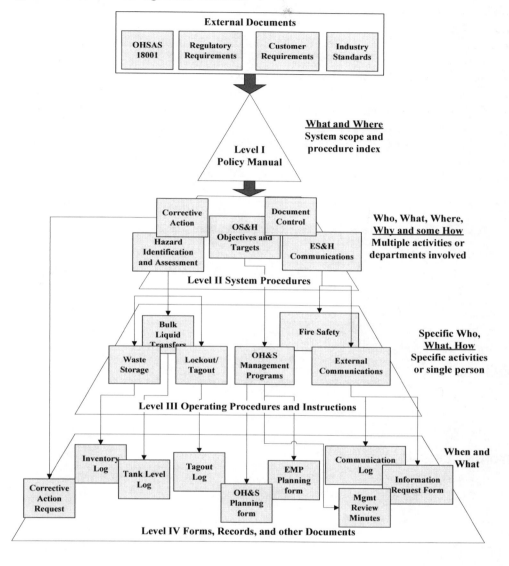

Instructions, also called level-three documents, provide detailed information on how to complete a specific task. A company may have a level-two lockout-tagout procedure, which provides common rules for all types of lockouts and many level-three lockout-tagout instructions that provide details on the steps needed to properly isolate specific components or circuits. Whether to use one large procedure or many smaller and more focused instructions is a decision that the organization must make. The best advice is to get the opinion of the users of the document. What would be most convenient for them?

Finally, level four includes all those forms (which will become records), lists, and other types of documents that need to be controlled. The pyramid structure implies that, in general, there are more documents of each type as one moves down the pyramid and the level of detail provided in each type of document increases as one moves down toward the base.

Although the structure just described is the most common, keep in mind that the structure of an organization's document control system is up to them. An organization could elect to put everything in one procedure if they wanted, although such a structure would be cumbersome to use and is not recommended. The bottom line is that the system an organization elects to establish must work, as evidenced by employee's familiarity, access to, and use of the documentation.

The exact format of an organization's documentation is likewise up to them. Many companies favor using flowcharts, some prefer textual, while still others prefer a combination of text and flowchart. A recommended format that has worked well is included on the CD. In this format there is a cover page that includes general boilerplate information (shown in figure 5.3), a flowchart that shows the essential flow and major steps in the process, and a textual portion that provides a place to expand on any details that cannot be put into a flowchart. This hybrid approach is probably the most flexible.

Note that the boilerplate includes sections that detail key inputs, key outputs, and metrics. It is recommended that an organization includes this key information for all of level-two (i.e., system-level) OH&S procedures. It is not normally necessary to include this information on instructions and forms. This format

Figure 5.3 Example Procedure Format: Health and Safety Planning

1.0 PURPOSE

This procedure defines the process used to determine significant workplace hazards and associated risks, methods for controlling and reducing the risks of these hazards, and the process used to determine and achieve health and safety objectives.

2.0 SCOPE

This procedure covers all areas and operations within the WeProvideIt, Saline, Michigan, facility.

3.0 DEFINITIONS

Job Hazard Analysis (JHA): process for identifying where hazards exist and determining the nature of the hazard.

Objective: goal set for OH&S performance.

Management Program: the means, time frames, and personnel responsible for achieving an OH&S objective.

4.0 RESPONSIBILITY

The Management Appointee has the overall responsibility for ensuring that targeted job hazard analysis and risk assessments are performed as scheduled and that workplace hazards, objectives, and programs are maintained up-to-date. The Safety Committee has the responsibility for review and approval of all JHA and risk reduction activities. Area supervisors are responsible for timely completion and update of JHA and risk assessments.

5.0 KEY INPUTS

- Health and Safety policy commitments and criteria for risk assessment
- Health and Safety regulations and other voluntary requirements
- Area Job Lists
- Information on facility, process and process changes and modifications

6.0 PRIMARY OUTPUTS

- Identification of the workplace hazards and associated risks
- Identification of methods to control these hazards and reduce their risks
- Identification of methods for monitoring the effectiveness of risk reduction activities
- Identification of OH&S objectives and management programs, and the achievement of these objectives

7.0 METRICS

- Monitoring of the effectiveness of this process will be through the monthly safety committee reviews, external audit surveillances, safety performance monitoring, and by the success in achieving defined OH&S objectives.

borrows from the "process approach" used in quality management and recognizes that for the process to deliver results it is not enough to simply follow the steps of the procedure. Instead, the key inputs into the process must also be available and controlled or the old adage "garbage in, garbage out" will prove true. Likewise, the key outputs from the process must be known and determined to be acceptable, because these outputs will in fact be inputs into other processes. The metrics section details the methods that will be used to ensure that the process is producing good outputs and achieving planned results.

Records are one type of document and must be controlled as documents before they are filled out and then as records once the information on them is recorded. Control of records is especially important in an OH&S management system, because many of the records have regulatory retention periods. Many regulatory records have a five- or seven-year retention period. Some must be retained far longer, up to thirty years. An organization must check the regulation to confirm the mandatory retention period. Nonregulatory records should be kept based on their potential need. Typically records for management reviews, internal audits, and corrective action are kept for three years to give the company a solid history of problems and improvements. Remember that the primary use of a record, beyond demonstration of compliance, is in analysis for improvement. Records provide the information needed to establish trends and patterns and should be used for same.

The OHSAS 18001 specification requires that an organization establish retention periods for records. The retention period may be listed as a minimum retention period, which allows an organization to keep them forever, if they so choose. Most experts caution against this, however, as there is a cost, in terms of money, effort, and space, associated with keeping records forever. It is best to dispose of records after they have exceeded their specified retention periods, typically through some type of annual review and disposal program.

An organization must also ensure that it has identified where the records are stored to permit easy identification and retrieval. A clear and consistent indexing method to quick location of the specific record needed should also be established. By indexing method I mean by chronological date, by activity, by report number, or some combination of these. It is especially important to clearly label boxes or (for electronic storage) file folders with their contents. An organization must also ensure that the locations used to store records safeguard them from damage, deterioration, and loss (or inadvertent deletion).

What records must an organization keep? Certainly all legal records required by environmental regulations. These include accident records and reports, reports of injuries, results of health examinations (e.g., audiograms for hearing), and certain training records, to name but a few. Also all records mandated by the OHSAS 18001 specification must be retained. Examples of these include records of management reviews, audit results, and records of compliance reviews. Beyond that, an organization must keep records that demonstrate that it is complying with its HSMS. As a general rule, an organization should keep records of all formal reviews and approvals along with records of all ongoing OH&S monitoring.

An example of a document and record control procedure is provided on the CD provided with the book.

DOCUMENTATION DEVELOPMENT

It is important that any documentation developed to ensure the safety and health of the organization's employees not only be well structured and contains the necessary information but also serve the needs of the document user. A common complaint associated with documentation is that it is never used; in most cases this is because of a deficient document more that a refusal on the part of potential users to read them. Some tips for safety documentation include:

- Only include what is truly necessary in the procedure. Unnecessary information makes the procedure cumbersome to use and detracts from the important information.
- Highly critical steps within a procedure should be boldfaced or otherwise highlighted.
- Number steps in the sequence in which they should be performed (e.g., 1.0, 2.0, 3.0). Actions within

a step (e.g., 1.1, 1.2, and 1.3) can be performed in any order. This methodology for numbering steps allows for both discipline and control where needed (i.e., steps must be performed in the sequence indicated) and flexibility (i.e., substeps within a main step can be performed in any order) where appropriate.

- Try to maintain a common format for procedures and instructions, where possible. Users should know where to find the information they seek without having to search through the entire procedure. Different types of procedures (e.g., maintenance procedures, operating instructions, and lockout procedures) may have differing formats, but all documents within the general category should use the same format.
- Ensure operating instructions contain or reference other procedures that include the instructions for the proper operation of equipment and systems. Include prestart-up checks, valve, switch, or electrical lineups, and instructions on how to secure the equipment.
- Include appropriate safety instructions including calling out operational controls (e.g., required PPE) in the procedures. Prominently highlight warnings and precautions. Include emergency shutdown instructions where appropriate.
- For critical or especially hazardous operations, include training or qualification requirements.
- For nonroutine or especially critical or hazardous operations, require operators to "cookbook" the procedure. Cookbooking means obtaining the procedure and performing it step-by-step as called out in the document. Do not allow operators to rely on memory for these tasks.
- Consider establishing a requirement that all process operators signoff on the back of the instruction to signify that they have read and understand the procedure, including any safeguards or PPE required to perform the job safely. Include a statement on the back page that states: *"Signature indicates that the operator has read and understands the procedure and his or her responsibility for safe operation of the process."* This provides a good record of training and provides a psychological driver to understand and adhere to it. The same action should be taken for JHAs if posted at the job site.
- Finally, use process operators to review procedures and instructions before issuance. Consideration should be given to including team leaders, supervisors, or even operators in the approval chain for procedures and instructions. The user of the document is in the best position to know if it will meet his or her needs. The users will do a credible job if you include them in the formal signoff. Most people are consciences of things they put their signature on.

COMMUNICATIONS AND CONSULTATION PROCESS

There are two elements of the communication process that must be addressed. The first is a system for receiving and responding to outside inquiries. The second is a system of internal communications to communicate important information about the HSMS to the workforce.

To establish a system for dealing with outside communications an organization needs to do two things. First, decide how such inquiries and other communications will be routed. Who will handle a request for information about the HSMS or safety performance? Because OH&S management does occur within a framework of regulatory control, most companies designate one individual as the focal point for all questions or issues dealing with the HSMS, including questions dealing with accidents and injuries. For most companies that individual is the management appointee. In some companies the responsibility for responding to such inquiries is distributed. The public affairs officer might respond to general questions from the media, the corporate attorney or safety engineer to any questions relating to regulatory compliance, and the management appointee to other requests for information. Even if the responsibilities are distributed there still should be a central focal point to log in the communication and to assign it to the right person. This leads us to the second element of the external communication program, the communication log.

In essence a communication log is a simple form used to document the nature of the communication, who it was from, the date, who it was assigned to, and the response provided. The OHSAS 18001 specification requires a procedure or process for external communications (i.e., communication from "interested parties"). An organization will need to show evidence of compliance, and while not required, a communication log is one of the easiest ways to do that. An example log is shown in figure 5.4.

The log would be maintained by the individual designated as the central focal point for external communications. Agreement should be reached on the types of calls that should be logged. An organization also needs to make sure that the main receptionist knows who this person is so that he or she can route the calls to this individual. A copy of the communication log is included in the planning workbook on the CD.

The second element of communication involves internal communications. The company first needs to

Figure 5.4   Health and Safety Management System Communication Log

COMMUNICATION LOG						
Date	**Name**	**Contact Info.**	**Communication**	**Assigned to**	**Response**	**Date**
1/15/2006	C. Tomlins - Carriers Insurance	734-944-9834	Requested information about our OHSAS 18001 status.	T. Baker - Mgmt Appointee	Faxed copy of our 18001 certificate and OH&S policy	1/15/2006
1/25/2006	MIOSHA - T. Bennet	763-934-0583	Called to provide information regarding the VPP technical visit.	T. Baker - Mgmt Appointee	None required	NA
4/12/2006	WDXV TV - Sheryl Weis	964-837-8976	Indicated wanted to do a local segment on our 18001 system and how it has helped us improve our safety performance.	K. Pierce - President	Pending	

consider what information needs to be communicated. Then methods to provide (or collect) this information can be developed. I normally recommend that the company walk through the OHSAS 18001 specification, picking out information that must be communicated, and then assign communication methods for each item. An example of a table created to accomplish this is shown in figure 5.5.

The table shows some of the information items that will need to be communicated. Undoubtedly an organization will uncover others as it develops its HSMS. The point of the table is to systematically analyze its communication needs and then to design programs for communicating relevant information throughout the workforce. Some of the methods in the table have already been discussed; others will be discussed in the sections that follow.

Remember that communication is more than a requirement; it is also an organization's main tool for building employee support and involvement. And as noted in chapter 2, significant, sustained improvement will not be possible without everyone's active support and involvement. Spend the time and effort to build effective two-way communication systems.

The final element of this system is employee consultation. As noted in chapter 2, the 18001 specification requires that employees be involved in the development and review of policies and procedures, be consulted regarding changes that could affect workplace health and safety, be represented on health and safety matters, and be informed of their representatives on OH&S matters, including the management appointee. The information components of these requirements should be addressed via the communication methods shown in figure 5.5. This section considers employee involvement and consultation regarding procedures, policies, and changes.

As previously noted the 18001 specification is not prescriptive. An organization can implement any structure that achieves the intent of the preceding requirements. Having said that, in most organizations, clearly the best path to fulfilling the requirements for consultation belong to the safety committee. Most organizations have a standing safety committee made up of representation from all functional areas of the organization, including both hourly and salaried personnel. As such, this committee is a natural choice for providing representation for the workforce. Because the safety committee's responsibilities include promotion of the safety and health of the organization's workers, involvement in the development and review of policies and procedures aligns well with both their assigned responsibilities and the requirements contained in 18001. It is recommended that an organization use the safety committee as the primary method for meeting the consultation requirements of OHSAS 18001.

The safety committee's charter should be examined and any necessary modifications should be made to ensure that the following responsibilities are known and communicated to the entire workforce:

- review of JHA and risk assessments,
- review of policies and procedures used to reduce workplace risks,
- representation of all employees on matters relating to workplace health and safety,
- communication with the workforce on issues relating to workplace health and safety, including changes that could impact the safety of employees,
- promotion of safety awareness within their work areas,
- assistance in accident investigation and review of the adequacy of the response, including any emergency response procedures,
- participation in routine safety inspections, and
- participation in any safety awards or campaigns.

In addition, if the primary responsibility for conducting JHA and risk assessment is given to the safety committee then this, too, should be identified in the charter. If the organization has no charter or equivalent document laying out the safety committee's responsibilities then one should be developed and reviewed with

Figure 5.5   Examples of Communication Methods for Occupational Health and Safety Management System Information

Clause	Communication Item	Methods for Communication
4.2	Health and Safety policy statement—internal communication of	Banners, newsletter, orientation training, local posting in each work area.
	Health and Safety policy statement—external communication to public	Posting in lobby, company website.
4.3.1 4.4.2	Significant workplace hazards—awareness of	Posting of final JHAs in each work area, orientation training, on-the-job training.
4.3.2	Regulatory requirements affecting normal operations	Embedding requirements in local procedures and instructions.
4.3.3 4.3.4	Health and Safety objectives and associated management programs	Business plan, posting of OH&S objectives on communication boards, company newsletter.
	Health and Safety objectives—obtaining the views of interested parties	Internal—gather concerns, issues and suggestions from the employee suggestion program and safety committee representatives. External—communication log.
4.4.1	Roles, responsibilities and authorities	Job descriptions, organization chart, embed in procedures and instructions.
4.4.2	Competence and Training	Training requirements matrix, training plans, on-the-job training.
	Awareness training—importance of conformity with policies and procedures, consequences of not following procedures	Orientation training, on-the-job training, company newsletter, JHAs posted in each area.
4.4.3	Representation on OH&S matters	Orientation training, postings.
4.4.4	Documentation and Document Control—location of and revision status	Master list available on the P drive (orientation training will demonstrate how to access and use the master list).
4.4.6	Operational controls	JHAs posted in each area, on-the-job training, embed in operating procedures and instructions.
	Health and Safety hazards of purchased goods and materials	MSDS sheets maintained in shipping and receiving and production supervisor office. Safety labels on purchased equipment.
	Health and Safety hazards of service providers and contractors	Obtain information through use of the Contractor Briefing Form provided to each prospective contractor.
	Your company's relevant OH&S policies and procedures—communication to suppliers and contractors	Incorporated into Supplier Quality Manual provided to each supplier and contractor. Also part of Contractor Briefing Form.
4.4.7	Emergency Preparedness and Response procedures and actions	Orientation training, drills.
4.5.1	Health and Safety monitoring requirements and results	JHAs posted in each area, embed in local procedures and instructions, mandatory management review of all compliance review reports, semiannual management review, monthly safety meetings.
4.5.2	Nonconformities identified by employees and reviews of accidents, incidents	Corrective action requests which can be initiated by any employee, incident review signoff by senior management.
4.5.4	Audit results	Mandatory distribution to senior management.

the committee to ensure compliance with the requirements of 18001 to define, document, and communicate roles and responsibilities relating to health and safety management.

One final word of advice regarding safety committees: Inclusion of a domineering senior manager, especially one who is more interested in production than worker safety, can single handedly destroy the effectiveness of the safety committee. An organization should choose management representation carefully.

Vendor and Subcontractor Control

Suppliers are an important part of an organization's HSMS. We will discuss two categories of suppliers and two primary requirements. The two categories are suppliers of materials and products and suppliers of services. The two requirements are related to obtaining information about the workplace hazards of supplied materials and services from the supplier and communicating relevant information about an organization's HSMS and policies to these same suppliers.

For providers of materials and parts, information regarding the hazards and potential impacts of their products is normally obtained through receipt of a material safety data sheet (MSDS). The MSDS provides information on the hazards associated with the product along with any precautions necessary for its safe use or disposal. It will also normally note any regulations that apply that can help categorize the material as to its hazardous or nonhazardous nature. The biggest problem with the MSDS is that the format may vary. Some provide detailed information, others do not. Learning how to read the MSDS is an important skill for the health and safety professional. The MSDS must be available to the organization's employees as part of the employee right-to-know laws. It is a good idea to establish a policy whereby a safety representative is given an opportunity to review the MSDS for any new chemicals or products purchased by the organization so that the materials can be assessed for their health and safety effects and so that operational controls can be established to reduce the risk to the organization's employees of exposure before the use of these materials.

Health and safety requirements are normally communicated to material providers as part of the contracting process. These requirements include prohibited or restricted materials and any requirements for certification or testing of materials.

Service providers are a little more complicated, because the nature of the service may vary considerably. Examples of service providers who need to be considered include:

- facility maintenance providers,
- bulk material providers (hazardous chemicals or materials, etc.),
- construction firms (expansion, renovation, demolition),
- laboratory services (for safety system monitoring equipment calibration), and
- professional services firms (i.e., safety consultants).

A company has an obligation to understand what hazards may be created as a result of the contractor's work and to ensure that it is properly controlled while the firm is working on its behalf. An organization must also communicate relevant information relating to its health and safety policies and procedures that these firms need to be aware of during their work. This may include policies for securing of equipment, lockout-tagout, bulk liquid unloading, welding and fire watches, emergency response, and confined space entry and control. The most common way to both obtain information from and communicate information to the supplier or contractor is through the use of a contractor briefing form, a portion of which is provided in figure 5.6. Note that this form typically includes information related to both environmental and health and safety control and management.

Other sections in the briefing form list the organization's policies and procedures that the contractor will

Figure 5.6   Contractor Briefing Form

Health and Safety Briefing Packet and Contractor Method Statement Template

Section V. Contractor Method Statement:
Respond to the following questions: (use additional space where required)

This method statement must be completed, signed, and returned to the facility's Health and Safety Manager before contracted work commences.

Work Description:
Briefly describe the work to be performed while on-site including the activities of subcontractors.

_____

_____

Electrical Safety
Will you be working on or require access to electrical components?   Yes ____ No ____
If YES, list these components and your method for preventing accidents or injury.

_____

_____

Working Aloft
Will the work you perform require working at heights greater than 8 feet?   Yes ____ No ____
If YES, how will the safety of your employees be safeguarded? Also list measures that will be taken to prevent injuries from falling objects.

_____

_____

Materials
What materials (chemicals, oils, solvents, etc.) and/or equipment will you be handling or bringing on-site to perform the contracted work? How will they be disposed of?

_____

_____

need to comply with during the work, a listing of licenses, or certifications held by the contractor, key names and contact data, and information relating to emergency response. A copy of the contractor briefing statement is provided on the CD.

Although it is not required by OHSAS 18001, it is not a bad idea to set up vendor history files for an organization's key contractors or suppliers. These files provide a place where an organization can store the contracts, contractor briefing forms, and details of any problems it may have had with the contractor or accidents or incidents they may have caused or been involved in. This history should be reviewed before extending or offering a new contract with the supplier. Careful evaluation and selection of service providers is the first step toward improving safety performance for activities that involve contractors or suppliers.

Calibration Systems

Health and safety monitoring equipment includes items such as audiometers, gas analyzers, and alarm systems associated with equipment and storage tanks. This equipment in effect serves as operational controls to ensure the safety of the workplace environment and an organization's employees. If a company has such equipment, and many companies do, then it must ensure that the equipment is reliable and accurate. Periodic calibration and maintenance of these devices is required to provide confidence in the data these instruments provide.

If a company already has a quality management system (QMS) in place then it is a simple matter to include these measurement devices in your existing calibration program. Most manufacturers, even without a formal QMS, recognize the need for calibration of measurement devices and will therefore have some type of calibration program in place. If not, an organization will have to develop a program. In many cases it will be easier to outsource the calibration of the organization's monitoring equipment rather than go to the expense of buying standards, obtaining calibration protocols, and training employees how to properly perform the calibrations.

At a minimum, an organization will need a list of all the measurement and monitoring devices that require calibration, their serial numbers or other unique identifiers, their calibration frequency, the date last calibrated, and when they are next due. For a simple system this information could be included as a sheet in the health and safety planning workbook. An example is shown in figure 5.7. For a more complex system with many devices, a dedicated calibration database is normally used. An organization may also want to list any periodic maintenance or checks performed on monitoring and alarm systems. Finally, an organization will need to set up a system to store the records of calibration. Make sure someone reviews the results of the calibrations if this activity is outsourced to a laboratory.

An entire book could be devoted to establishing and maintaining a calibration program. We have provided the basic information here to get started. Again, the best option is to fold monitoring equipment into an existing calibration program if possible.

Training, Competency, and Awareness

There are two requirements that an organization's training program must meet to satisfy OHSAS 18001. The first is competency. All personnel whose work could affect workplace health and safety must be competent. Competency is more than just providing training. Competency attainment means that the individual has mastered the essential skills needed to safely perform the task. Competency may be attained through training, formal education, and experience, or more commonly through a combination of these.

There are three primary steps involved in establishing competency. The first is identifying who needs to be trained and what knowledge and skills these employees must have. Competency identification should have

Figure 5.7   Simple Calibration List

Calibration							
Device	Serial No.	Location	Freq.	Date Calibrated	Date Due	Maint.	Comments
Decibal Meter	A3482	Safety Coordinator Office	Annual	5/12/2006	5/12/2007	Reference check prior to use	
Temperature Sensor	T174	Plating tank 1	Annual	3/15/2006	3/15/2007		High Temperature Alarm sensor for plating operation
	T175	Plating tank 2	Annual	3/15/2006	3/15/2007		
	T176	Plating tank 3	Annual	3/15/2006	3/15/2007	None	

already started as part of the JHA. The CFT performing the review was asked to note the functions involved in the activities associated with the identified task and therefore the job hazards. This information was recorded on the health and safety planning workbook or the JHA form. This information provides a good starting point for identifying the competencies that will need to be achieved or verified during training program development.

In addition, regulatory agencies also stipulate certain required training, depending on the activities an organization performs and the nature of risks involved. An organization must review the applicable regulations to see which of these items apply to them. A partial listing of some of the training that may be required includes:

- Hazardous waste operations and emergency response (HAZWOPER),
- Blood-borne pathogens,
- Use of PPE,
- Respiratory protective equipment,
- Fire protection,
- Confined space entry,
- Lockout-tagout,
- Emergency action plans,
- Hazard communications,
- Chemical hygiene,
- Process safety management, and
- Hazardous material handling.

Each of these training items has specific applicability requirements and training topics that must be included. In addition, most have refresher training requirements that must be adhered to. OSHA provides a booklet, number 2254 titled *Training Requirements in OSHA Standard and Training Guidelines*, that can help. A copy is included on the CD.

The focus of the training should be on the hazards, risks, and controls associated with each job. These items can be shown on a worksheet of the health and safety planning workbook, an example section of which is shown in figure 5.8.

In this matrix the competencies are listed along the rows. These are separated into two components—the job specific competencies and the general competencies—required for all employees. The training elements focus on the procedures used to control the activities and in some cases demonstrations of the competencies by performance. Each employee in the department is listed across the top, and the functions the employee fills are shown beneath their name. The required competencies are shown with an *X* in the cell formed by the competency row and individual's column. An *X* indicates that the competency has been assigned but not yet attained. An actual date replaces the *X* when a competency is achieved or verified. The *PE* in the last column for the manager indicates that this individual was "grandfathered" in based on previous experience. A simple spreadsheet calculation shows the percent competencies achieved, which can be useful for training performance tracking.

OHSAS 18001 also requires awareness training. Competency involves knowing *what* to do and *how* to do it, whereas awareness implies an understanding of *why* it is important. The specification requires that all employees have an awareness of the following:

- The importance of complying with policies, procedures, and the requirements of the HSMS.
- The significant workplace hazards they are involved in along with the risks and benefits of improved personal performance.

Figure 5.8 Identifying Training Topics

Department/Function: Material Handling — Personnel			Robertson	Lindel	Tomy	Sizemore	B. Smith	Dillon
Functions Filled								
		Manager						X
		Supervisor					X	
		Material Handler	X	X	X	X		
Job Specific Competencies								
Hazard	Competencies	Training Elements	Competency Requirements					
DynaFoam	Handling, storage and disposal	Material Handling Procedure	6/1/2006	X	6/8/2006	6/8/2006	5/10/2006	PE
DynaClean 2000		Waste Handling and Disposal Procedure	6/1/2006	X	6/8/2006	6/8/2006	5/10/2006	PE
HiLo Batteries		Bulk Liquid Transfer Procedure	5/11/2006		X		5/2/2006	PE
		Perform bulk liquid transfer	5/15/2006		X		5/7/2006	
HiLo Batteries	Battery Charging	Battery Charging Instruction	6/1/2006	X	6/8/2006	6/8/2006	5/10/2006	PE
HiLo Accidents	Proper Operation of HiLo	Certification	4/6/2003	4/13/2006	12/1/2004	4/13/2006	9/5/2004	
		Annual Refresher	3/31/2006	4/13/2006	3/31/2006	4/13/2006	3/31/2006	
General Competencies								
Fire, Explosion, workplace violence	Knowledge of evacuation process	Fire and Evacuation Instruction	5/20/2006	X	5/20/2006	5/20/2006	5/20/2006	5/20/2006
Fly objects, noise	use of personal protection equipment	Job Safety Handbook	5/20/2006	X	5/20/2006	5/20/2006	5/20/2006	5/20/2006
		Competencies outstanding	7					
		Competencies achieved	36					
		Competencies based on PE	4					
		Total Competencies Assigned	47					
		Percent Competencies Achieved	0.85					

- Their individual roles and responsibilities (including those relating to emergency response).
- The potential consequences of failing to follow specified operating conditions.

This awareness can be provided through orientation training, on-the-job training (OJT), the company newsletter, all-hands meetings, or a combination of these. Note that the specification indicates that the training process take into account differing levels of responsibility, ability, literacy, and risk. It is strongly recommended that area supervisors be given the specific responsibility to follow up on, if not personally provide, awareness training to area personnel.

One of the best tools to help in maintaining a high level of awareness is by posting the completed JHA form at the work site. This simple form provides a summary of the job hazards, the risks associated with those hazards, and the controls used to minimize those risks. It can also be used by area supervisors to guide their delivery of awareness training to area personnel.

Other methods for maintaining a high level of awareness among plant staff include:

- Large safety posters and billboards displayed prominently throughout the plant. They are particularly effective at employee entrances.
- Distribution of safety pamphlets and folders, commonly available free or at minimal cost from government agencies.
- The establishment of safety communication boards that include areas of current focus, annual safety objectives, articles of safety interest from newspapers and other sources, safety team pictures and names of area safety representatives, before and after pictures of safety improvements, and posting of MSDSs for new chemicals or materials that also include safety metrics, including plant and area long-term

safety performance and a Pareto chart showing the frequency of occurrence of different types of injuries (e.g., slips, strains, cuts, burns, etc.) over the last period of time, typically monthly or quarterly.

The final step in establishing an effective training program is to verify the effectiveness of the training or other actions taken to establish the competencies and awareness needed. Methods of training verification include interviews, tests, personal observation of the employee performing the task, and review of data relating to how well the activity is being performed. In addition internal audits can and should be used to verify competency and awareness. Internal auditing will be discussed in final section of this chapter and in chapter 8.

Finally, keep in mind that an organization must keep records of its training activities. Some of these records are required by law; others are needed to provide evidence that an organization's HSMS is operating as it should. Records may be in the form of certificates of successful completion of a training course, sign-in sheets providing evidence of attendance, and supervisory signoff for OJT activities. While keeping a formal record of all training that occurs on a daily basis is impractical, an organization should ensure they have records for all regulatory required training, all formal training, and all initial training. One innovative and simple way to show evidence of training to organizational procedures is to have the employee sign the back of the procedure when training is completed. The procedure, which should be available at its point of use, then provides an easy way to verify who has and who has not been trained. When a procedure is revised and the new procedure replaces the superseded one, then this method ensures that all team members are trained on the new procedure.

Accidents, Incidents, Nonconformance, and Corrective and Preventive Action Systems

Systems must be in place to properly respond to accidents and incidents in operations and to weaknesses in the occupational HSMS. Before reviewing the steps necessary to develop such a system it is worthwhile to discuss the differences between corrective action, preventive action, and continual improvement.

Corrective action is action to address an actual violation to requirements. Requirements in this sense include violations of regulatory requirements as well as violations to what an organization says and does in its procedures and policies that make up the HSMS. These occurrences represent nonconformities. Preventive action is action to address a potential violation to requirements before one actually occurs. Both corrective and preventive actions seek to identify the root cause of such violations and to put in place effective countermeasures to eliminate the cause and prevent the occurrence (preventive action) or recurrence (corrective action) of the nonconformity. An example of a corrective action would be actions taken to identify the root cause of a spill of acid in a laboratory area. An example of preventive action would be the recognition of the potential for such a spill before one happens and taking action to correct the practices that could lead to the incident.

Continual improvement, on the other hand, is action taken to improve a situation or performance that is not directly tied into a requirement. Continual improvement is best accomplished through the development of OH&S objectives and through actions to eliminate hazards in the workplace. Actions to eliminate a hazard, as opposed to controlling it and to establishing an employee wellness program go beyond regulatory requirements. Actions taken to install a machine guard over a potential pinch point before an injury occurs is more of an example of preventive action, although it does improve an organization's overall safety posture.

To establish a system for identifying, documenting, evaluating, and tracking nonconformities and resultant actions an organization will need to take the following steps:

1. Develop a form to document the nonconformities and any resultant corrective or preventive action.
2. Develop a process for determining root cause.

3. Develop a system for tracking the actions taken to identify and eliminate the cause.
4. Develop a system process for verifying the effectiveness of the actions taken (did they eliminate the cause?).

Forms used to document nonconformities and actions come in all shapes, sizes, and names. Common names include the corrective action request (CAR), preventive action request (PAR), action request, or nonconformance report (NCR). It does not really matter what it is called. If an organization has an existing QMS, it may be useful to borrow the same form used for quality issues. In this case some companies add an *S* for safety to the prefix for the form (i.e., SCAR, SPAR, or SAR). The choice is up to the organization. What is on the report is more important than its name. An example of an action request form is shown in figure 5.9.

Note that the action request form is not a replacement for an incident or accident report generated after the actual occurrence of an injury or incident. It is rather a tool that provides for the systematic identification of root cause and short- and long-term actions needed to mitigate the effects of the problem and to eliminate the cause so it will not happen again.

The form shown in figure 5.9 could be used for any type of system deficiency, audit finding, or other event requiring root cause analysis and permanent corrective action. It shows the immediate containment action, root cause analysis using 5 Why and Cause and Effect Analysis, permanent corrective action, verification of action, and horizontal application (what other activities would benefit from the permanent actions?). It also documents the results of risk assessment of the proposed actions, which is required by 18001, to ensure the cure is not worse than the disease. Finally, checkboxes to show document updates have been added to ensure that no supporting actions are overlooked. This format, which goes well beyond that required by OHSAS 18001, is borrowed from the QMS and is common to the automotive industry.

To ensure the problem does not recur, some method of root cause analysis is needed. Some of the more common tools are discussed here. Note that these are summary descriptions only. Skillful application of these techniques takes practice and maybe some training.

- *Brainstorming*: In brainstorming team members are all asked to come up with as many ideas on potential causes as they can. A team of five to seven people works best. Instructions should include:
  - no criticism of ideas is allowed,
  - the goal is quantity of ideas not necessarily quality of ideas at this point,
  - "outside the box" ideas are welcomed,
  - team members may pass on a round if necessary, and
  - build off of other's ideas.

  All ideas should be recorded on the flipcharts. Assign one or two team members to collect ideas. Continue until the team runs out of ideas. Allow humor; it will result in team members feeling more comfortable and contributing more ideas. A typical brainstorming session will generate between fifty and one hundred ideas and up to two hundred ideas is not uncommon.
- 5 Why Analysis: this technique drills down to the root cause by continually asking the question why after each answer. One is essentially looking for two types of root cause—the technical or behavioral root cause (e.g., no one verified the valve lineup) and the system root cause (e.g., there is no requirement to perform a valve lineup after sump draining or prior to refilling). Sometimes it takes only four "whys" to get to the root cause, sometimes six.
- Cause and Effect (Fishbone Diagrams): this technique is essentially guided brainstorming. It is guided because the team will systematically consider each main category shown on the "fishbone" and ask the question "Is there any way that this category could have caused the problem?" In our example the categories are Methods, Human, Health and Safety, Materials, Equipment, and Tooling. These are the typical default categories. One can then show the main results on the fishbone diagram embedded in the action request. In practice the analysis would be done on a whiteboard or flipchart with the team,

Figure 5.9   Action Request Form

Action Request Form			Source		Type	
Problem ID: 12-06	Team Contact:	Gyuna	Audit	☐	Repeat	☑
Date: 6/14/2006	Champion:	B. Smith	OH&S	☑	Corrective	☑
Area: Production	Activity/Process	Maintenance	EMS	☐	Preventive	☐

Problem Description	Problem Impact:	1 hour lost productivity	Response Due Date:	6/20/2006

The control circuit for the furnace was found to be still energized during a routine circuit check prior to maintenance even though the power had been locked out. This is the second occurrence of an energized power circuit after lockout this year.

Immediate Containment Action	Who	Date
The panel was immediately secured	B. Smith	6/14/2006
The power to the control circuit was secured and locked out	B. Smith	6/14/2006

Problem Root Cause Analysis

5 Why Analysis

1 Why: The circuit was energized because the fuse to the circuit was not pulled.

2 Why: The fuse was not pulled because it was not identified in the lockout procedure

3 Why: The circuit was recently modified and the fuse added but the procedure was not updated

4 Why: There is no requirement to update lockout procedures after equipment modifications

5 Why

Root Cause: there is no requirement to update lockout procedures after equipment modifications

Cause and Effect Analysis

Methods   Human   Environmental

Materials   Equipment/Tooling

Error Proofing   Was error proofing a part of the long-term actions?   Yes   No   ☐   ☑

Long-Term Actions and Verification		Date	Indicator	Before	After
Action Who	B. Smith - Update the lockout procedure for the furnace control circuit	6/19/2006	Procedure	Fuse not indicated	Fuse indicated
Action Who	B. Smith - Add a new requirement in the Management of Change procedure to require review of lockout procedures	6/19/2006	Procedure	No rqmt.	Rqmt. In procedure
Action Who	Gyuna - Train all maintenance and engineering personel on the new changes.	6/20/2006	Training records	No training	Training complete
Acton Who	Tomlins - Complete internal incident critique	6/15/2006	Lockout incidents	2 last year	TBD
Action Who	Tomlins - Complete a review of all other modifications over the past 3 years and verify lockout procedures	6/30/2006	Review	No review	
Action Who	Tomlins - Complete any required modifications to other lockout procedures based on results of review	7/30/2006	List of revisions	Not updated	

Systematic Prevention and Horizontal Application

All lockout procedures for any equipment that has been modified over the past 3 years will be reviewed and modifications made as needed.

Risk Assessment of Proposed Actions Completed?	Yes ☑	No ☐	No.

No additional risks were identified as a result of taking these actions.

System Updates (Check those updated)		ECI/ECR Submitted?	Yes ☑	No ☐	No.	

MQC/IS	☐	MCCS/PM Instructions	☐	Standard Work	☐	Lockout Instructions	☑
JHA	☑	Inspection Sheets	☐	Training Plans	☐	Aspects Worksheet	☐

and there would be many more potential causes. The drawback is that the root cause is not as readily evident as it is with the 5 Why Analysis.

- Nominal Group Technique: If brainstorming or cause-and-effect analysis is used to generate possible causes then the team will have too many causes to act on. To get the causes down to a reasonable number, say five to ten for action assignment or to identify what the team feels is the primary root cause, some technique must be used to narrow down the list. Nominal group technique is one method to do this. It is a structured tool for ensuring that all participants have an equal voice in the decision-making process. A facilitator displays the causes to be ranked on flip chart. Each participant picks five items from the list and ranks each using a five-point scale. If there are a large number of items, the facilitator may instead have each member pick ten items and rank them on a ten-point scale. Typically the number ranked should be about one third of the number of items on the list. The votes from each team member are recorded on the flip chart next to the listed items and the summed score for each item is tabulated and the top items selected.

These are just a few of the tools that can be applied to help identify the root cause. The important point is that the organization should pick one or more of these techniques to use, master the technique, and then skillfully and consistently apply it to nonconformities as they are identified.

The next step is to develop some method of tracking the actions taken to correct nonconformities. Although you could simply track them using the long-term action and verification section of the action request form, this would be cumbersome if more than a few were outstanding and some actions would probably fall through the cracks. A simple spreadsheet or Microsoft Word table can be used to provide more comprehensive yet simple tracking of actions and follow-up. An example of a simple tracking log is shown in figure 5.10 and is included in the health and safety planning workbook provided on the CD.

The log, normally maintained by the health and safety manager or QMS representative, allows quick and easy tracking of the status of all action requests issued. Comments can be added to show the status of outstanding corrective actions. The OH&S manager or management representative should follow up on outstanding and overdue corrective action. He or she should also follow up to verify the actions were completed. The worksheet also has a section for summary statistics to tabulate performance in completing corrective action.

The last step is to follow up on the actions taken to ensure that they were taken and were effective in eliminating the cause and preventing recurrence. Verification of actions taken can be performed by sighting the changes made in procedures and policies, interviewing employees, or otherwise personally checking the results of the actions. Verification may also be performed as a part of HSMS internal audits by having the auditors confirm that corrective actions are in place when they next evaluate the area. Long-term verification of the effectiveness of the actions must normally wait some period of time to ensure the problem does not recur. The comments section of the action request log provides a convenient place to document verification actions. Also note that the long-term actions and verifications section of the action request form also has a section to document verification. The to be determined (TBD) in the "After" columns of the example indicate that the verification of long-term effectiveness will have to wait until several months have passed with no lockout-tagout errors before there can be confidence that the actions have solved the problem.

Internal Audit Process

The last core design element that we will discuss is the internal audit process. Detailed audit strategies will be discussed in chapter 8. This section discusses the general audit program. The company needs to have the capability to critically assess their OHSAS 18001 management system. To state in simple terms, the objectives of any management system audit is threefold:

Figure 5.10 Action Request Tracking Log

Action Request Tracking Log

AR Number	Date	Description	Repeat?	Action Assignee	Due Date	Action Complete Date	Verification Date	Comments	Status (R-Y-G)
008-06	4/14/2006	Internal audit found that several material handlers had not been trained on safe material and waste handling techniques	No	Harris	5/1/2006	4/27/2006	5/1/2006	Verified training records available, interviewed the 3 operators.	G
009-06	4/14/2006	Internal audit noted that Feb and March monthly PMs on the interlock circuit for the parts cleaning hood filtration system had not been performed	No	Kuhn	5/1/2006			Awaiting replacement switch	R
010-06	4/14/2006	Audit finding - No evidence of the completion of the Supplier Briefing Form for the new HVAC contractor could be provided.	No	Bennet	5/15/2006			Form has been sent out, no response yet. Bennet said she will follow-up 5/20/06.	R
011-06	5/3/2006	The safety check for the 50 ton overhead crane cannot be performed as written. The crane has been modified since the last performance and this has not been reflected in the PM.	No	Harris	5/18/2006	5/16/2006	5/16/2006	Verified procedure was corrected.	G
012-06	6/14/2006	Machine guard for the No. 2 rolling machine was not in place	Yes	B. Smith	6/20/2006			2nd occurrence this month.	Y
		Summary Statistics							
		Number Red	2						
		Percent Red	0.40						
		Number Yellow	1						
		Percent Yellow	0.20						
		Number Green	2						
		Percent Green	0.40						
		Total Action Requests	5						

- to identify any weaknesses that exist in the management system,
- to identify opportunities for improvement, and
- to identify best practices that should be communicated throughout the organization.

The extent to which these objectives can be met will depend on the maturity of the management system itself, the skill and training of the internal auditors, and the level of support provided by the management team for the realization of these objectives.

A fundamental purpose of any audit is to verify that the organization is conforming to the OHSAS 18001 specification and to its own internal policies and procedures. A management system cannot produce results if it is not followed. Noncompliance is a weakness that will lead to inferior performance of the HSMS, and such instances must be identified during the audit.

Conformance means that the organization is adhering to the requirements set forth in its internal procedures, policies, guidelines, and to external requirements set forth by the specification, regulatory agencies, or adopted industry practices. Basically, it verifies that an organization is doing what they say they do. Auditors generate NC findings to document areas where deviations to the requirements are noted.

OFIs are any areas where improvements in a process, typically associated with results, are obviously possible. An example might be an instance where an auditor notes that JHAs are not posted at the work site. The auditor is also told by the area supervisor that keeping the employees aware of the area's hazards is difficult. Note that the OHSAS 18001 specification does not require posting of the JHAs. Assuming the company also does not require posting of the JHAs, the auditor cannot generate an NC in this situation, because there is no definitive requirement being violated. Instead the auditor may note the issue as an OFI. An OFI does not require action on the part of an organization being audited, because there are not any defined requirements that must be met.

Identifying best practices is a valid objective of an audit, but one that requires significant perspective and objectiveness. Experienced internal auditors in particular are in a favorable position to identify best practices during audits so that they can be communicated and implemented throughout other portions of the organization, as appropriate. Citing a best practice is appropriate when the auditor finds performance to be significantly better than average and well beyond the minimum required to comply with the specification or the organization's basic requirements. The systematic identification of best practices, however, looks beyond the performance numbers and seeks to identify the enablers of performance so that others within the organization can judge whether the practice should be adopted by his or her team, department, or business unit. This type of "deeper dive" is primarily within the realm of internal auditors.

Verifying effectiveness is one of the most important objectives of any management system audit. This objective goes beyond basic correction of identified weaknesses and evidence of some improvement and focuses on the primary purpose of the process and the extent to which the process is accomplishing that purpose. As an example of **effectiveness confirmation**, assume that JHAs *are* required to be posted at the work site and that this is the primary method for maintaining employee awareness of work area hazards. The review for compliance would seek to verify that the JHAs are posted and are up to date. Confirmation of effectiveness would focus on whether they are effective in maintaining employee awareness. To verify effectiveness, the auditor would interview several area operators and question them on the hazards present. If they are not aware of area hazards then the system being used (posting of JHAs) is not effective, and in fact, represents waste because keeping the postings up to date requires time and effort with little results.

The basic elements of an internal audit program include the selection and training of the auditors, development of checklists to guide the auditors in what to look for, the development of an audit schedule, and the design of a system to track audit findings.

To be effective, internal auditors need to be trained both in the audit process and in the specification to which they will be auditing. Checklists are not required to be used for the internal audit, but they do help ensure a comprehensive review and consistency in audit performance. New auditors in particular will benefit from a detailed checklist. A set of audit checklists for OHSAS 18001 has been included on the CD. Note that these checklists combine related clauses together to form natural audit processes. Note that these checklists only address requirements contained in the OHSAS 18001 specification however. An organization must also ensure that it is complying with local policies and procedures.

The audit schedule shows the areas or topics that will be evaluated and when they will be conducted. A typical audit schedule covers one year. The schedule should be responsive to the needs of the organization. By responsive, I mean that the schedule should evaluate more important elements more often along with any areas where problems or issues have repeatedly occurred. For most systems this will mean that the deployment of the operational controls will be evaluated more often than will common system elements, such as document and record control. An example of a typical audit schedule is shown in figure 5.11. A template for this schedule has also been included in the health and safety planning workbook on the CD.

The schedule shows the areas to be audited. A pop-up comment in each topic's cell provides details on the scope of each audit. As in the other worksheets, a formula has been built in to calculate the performance statistics for audit completion.

Note that this audit schedule identifies the current OH&S objectives and what the organization believes are its most significant safety hazards. This provides guidance when scheduling the different audit topics. Those areas that are more important or which are vital to achieving the organization's objectives should be evaluated more frequently than other areas. The schedule also shows when the compliance reviews and drills are scheduled to be performed.

Figure 5.11 Example Health and Safety Management System Audit Schedule

Internal Audit Schedule - CY 2006	Most Important Safety Hazards											
Initiated - 12/16/05	Machine Guarding											
Last Revised - 4/16/06	Forklift Operation and Material Handling											
	Slips and Falls											
Audit Area/Process/Month	Jan	Feb	Mar	Apr	May	Jun	Jul	Aug	Sep	Oct	Nov	Dec
OH&S Planning				X		X						
Completed				04/12/06								
Status (R-Y-G)				G								
OH&S Improvement				X		X						
Completed				04/12/06								
Status (R-Y-G)				G								
Document and Record Control				X			X					
Completed				04/12/06								
Status (R-Y-G)				G								
Training, Purchasing, and Communication				X			X					
Completed				04/13/06								
Status (R-Y-G)				G								
Management Support				X					X			
Completed				04/13/06								
Status (R-Y-G)				G								
Operational Control - Production				X		X				X		
Completed				04/13/06								
Status (R-Y-G)				G								
Operational Control - Maintenance				X			X				X	
Completed				04/14/06								
Status (R-Y-G)				G								
Operational Control - Logistics				X			X					X
Completed				04/14/06								
Status (R-Y-G)				G								
Operational Control - Admin.			X									
Completed			03/09/06									
Status (R-Y-G)			G									
OH&S Compliance Review			X									
Completed			03/09/06									
Status (R-Y-G)			G									
Fire and Evacuation Drill					X					X		
Completed												
Status (R-Y-G)												

Number Red	0	Percent Red	0.00%
Number Yellow	0	Percent Yellow	0.00%
Number Green	10	Percent Green	100.00%
Total	10		

Health and safety compliance reviews are similar to an audit but are performed at a more detailed level and are normally done by those having specific knowledge of the regulatory requirements that the organization must adhere to. While internal management system auditors will check to ensure the compliance reviews are being completed, safety specialists normally conduct the compliance reviews. The compliance review will evaluate conformance to the detailed regulatory requirements that apply to the organization.

Many companies hire an outside consulting firm or use a corporate safety specialist to perform the review(s) if they do not have safety professionals on staff. In this case it is important that management receives a detailed report of the compliance review. An organization should be able to demonstrate that all regulatory requirements listed on its register of regulations have been examined and actions taken to correct any deficiencies.

Audit tracking systems normally mirror the system to track nonconformities and related corrective and preventive action. It is recommended, in fact, that the same system be used. Any nonconformities resulting from the audit should be written up on an action request form and tracked using the action request log. The same requirements for root cause identification, long-term action to prevent recurrence, and action verification should apply to audit NCs. Using one system to track all deficiencies is normally easier that maintaining separate systems.

Finally, note that records of the audits must be maintained. As a minimum, an organization should maintain copies of the completed audit checklists, the audit action requests, and the cover sheet that summarized the results of the audit. Most organizations keep these records for a minimum of three years to correspond to the typical certification cycle, which is also three years.

Summary

This chapter started with the identification of the organization's legal and other requirements and workplace hazards. Together with the policy statement, these actions form the core of the HSMS. Because of their importance, significant detail was provided in how to identify workplace hazards and assess their risks. The chapter then moved into the selection of appropriate controls to reduce risk, including the use of hazard elimination, engineered controls, administrative controls, and PPE. The design of common supporting systems, such as document and record control, training, communication, corrective and preventive action systems, and the internal audit program, was then addressed. Throughout this chapter tools and techniques were provided to facilitate the design of these systems. Especially useful is the health and safety planning workbook, which provides a central file that can be used to manage almost all elements of the HSMS. The workbook and the tools, along with example procedures, are provided on the CD provided with this book.

Although it has been noted in several sections, it is worth repeating that an organization should capitalize on any systems already in place as a result of prior implementation of a quality or occupational HSMS. This is especially true for the common support elements, such as document and record control, training, corrective and preventive action, and internal audits. While an organization can maintain completely independent systems if desired, it will take longer to implement and will probably result in some redundancy and possibly a little confusion if you do.

At this point the basic 18001 management system foundation has been designed, the initial round of JHAs conducted, and operational controls implemented. The next chapter examines additional issues involving the deployment of the system, including the development of a systemwide series of metrics for ongoing monitoring.

CHAPTER 6

Deploying the Management System

Keep in mind that design, development, and deployment is not a series of tasks, one after the other, but is rather a simultaneous process with multiple iterations. Operational controls to reduce risks should be deployed as soon as possible after completion of the JHAs. System procedures should be issued immediately after their development. This process is repeated many times until the complete system is in place. This chapter, then, will highlight some of the steps that must be considered during deployment, no matter when in that timeline that deployment takes place.

The major steps involved in deploying the HSMS include:

- the review, approval, and issuance of the systems policies and procedures;
- developing and then deploying monitoring and metrics;
- providing initial all-hands awareness training, if not already completed;
- providing activity-specific training on the procedures, policies, and other operational controls implemented; and
- conducting the first round of internal management system audits.

Review, Approval, and Issuance of System Policies and Procedures

Many of the company's operational controls will be in the form of policies and procedures. Identification of the needed procedures will have been accomplished during the design phase, and some procedures may have already been drafted. In this section we discuss some of the most common procedures needed to support a typical company's HSMS. Examples of these procedures are provided on the CD. We will also discuss the steps necessary to ensure the accuracy of these documents and to get them issued to locations where needed.

Each of these procedures, plus any others developed as part of the HSMS, should be thoroughly reviewed and approved by those having in-depth knowledge of the activity before issuance. Once approved, each procedure (or instruction) should be placed on the master list and then distributed to the locations where it will be needed or could be needed in the event of an incident. Where electronic document control is used, folders should be set up that provide a clear pathway to the document that any user can understand. Training of personnel will be required in the procedures and will be addressed in a later section.

Identifying Occupational Health and Safety System Metrics

The team also needs to define the measurements that will be used to monitor its performance in making its workplace healthier and safer. Almost all organizations with more than a few employees already have some safety metrics in place as a result of their legal obligations to report and communicate accidents and injuries. In the United States, OSHA's 300 and 301 forms do just this. These metrics only tell part of the story, and they are reactive in that they only report the results of the organization's efforts to improve health and safety and provide little information on the practices that led to the results.

Another weakness in putting too much emphasis on aggregate injury or lost workday statistics is that they can lead to counterproductive behavior that weakens the overall safety program. As an example, managers may reassign injured workers to temporary but less strenuous jobs to avoid having to record a lost workday injury. The overall lost workday results then become misleading. Of greater concern is the situation where because of a strong but poorly planned program to drive down accident rates, employees do not report minor injuries or near misses for fear of the punishment they will receive for not following standard safety practices. Even in programs where more positive reinforcement techniques are used, such as safety awards and safety competitions, a strong bias toward not reporting injuries can arise as employees seek to achieve the award (which may be a cash award or time off) or win the competition. None of these conditions will be captured by the metrics focused on the number of recordable injuries or lost workdays only, and in fact, these conditions can make such metrics almost counterproductive if they are the only measures of safety program effectiveness used.

What is needed is a **balanced scorecard** of safety metrics that report both the results of improvement activities and the activities and behaviors that drive results. The concept of balanced scorecards was popularized by David Niven and Robert Kaplan in their book *The Balanced Scorecard* and has found great popularity among senior management teams looking for ways to drive organizational performance improvement toward its strategic goals. I will borrow the concept of the balanced scorecard and apply it to health and safety programs in which the goal is to drive organizational performance improvement toward achievement of OH&S policy commitments and objectives.

A balanced scorecard seeks to balance the indicators used to measure organizational results against the drivers of that performance. In a HSMS, where behaviors can greatly affect the result indicators (i.e., injury rates, lost workdays, etc.) balancing the metrics of performance is of great importance. In such a system, the indicators of the drivers of performance have as much weight as the results indicators and receive the same level of scrutiny and action. An example of the balanced scorecard for an OH&S management system might appear as shown in figure 6.1.

The indicators of *performance results* and *stakeholder indicators* are reactive, in that they show performance over a past period but not necessarily how those results were achieved. Therefore, they lack predictive power in indicating what future performance will be. The other two, *internal process indicators* and *learning and growth indicators*, instead measure activities that drive performance and are considered leading indicators. Success in the activities that these indicators measure will drive performance results. Slippages in these indicators foretell disappointing results. They provide the basis for early intervention to help ensure that improved performance will be realized.

The rest of this section will expand on the scorecard and show how it can be used to measure overall health and safety performance. I will use the example policy statement from chapter 3 as the example for this discussion. This policy statement has been reproduced in figure 6.2.

To demonstrate progress in meeting these commitments, assume that the organization has established the following objectives for the coming year (note that the next chapter will go into more into the identification and deployment of OH&S objectives):

WPI OH&S Objectives for Calendar Year 2006

- Reduce the total number of reportable injuries by 50 percent from 2005's level.
- Complete JHA and risk assessments for all priority-one activities and for 50 percent of the priority-two activities as indicated on department job lists by the end of December 2006.
- Reduce the percentage of jobs rated high risk from the 2005 level of 38 percent to less than 25 percent by the end of December 2006.

Figure 6.1   Balanced Scorecard of Occupational Health and Safety Metrics

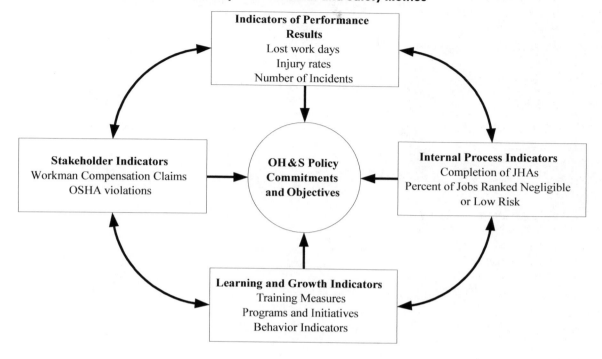

Figure 6.2   Example Occupational Health and Safety Policy Statement

WeProvideIt, Inc., is committed to the continued safety and health of its employees. In meeting this commitment, WPI will:

- Continually work to identify existing job hazards and to eliminate or reduce their risks to our employees.

- Comply with all applicable health and safety laws and regulations, and to go beyond compliance where needed to safeguard personnel.

- Comply with any other requirements needed to protect our employees, including participation in OSHA's Voluntary Protection Program.

- Seek to continually improve its health and safety performance through programs that focus on the long-term health and fitness of our employees.

- Establish a safety culture that includes management, salaried, and hourly colleagues working together to jointly ensure the safety of all colleagues.

- Achieve merit status in the OSHA VPP.
- Initiate a program to partially cover the expense of health club memberships.
- Pilot a behavior-based safety program in the assembly department.

Each of these objectives would then be measured during the year to ensure that the objectives were being met. A mapping of these objectives onto a balanced scorecard might appear as shown in figure 6.3.

Success in meeting the internal process objectives and learning and growth objectives should lead to success in meeting the indicators of performance results and stakeholder objectives of reducing the percentage of recordable injuries by 50 percent. By tracking these indicators month to month, the organization should be

Figure 6.3   Balanced Scorecard of Occupational Health and Safety Objectives, Indicators, and Initiatives

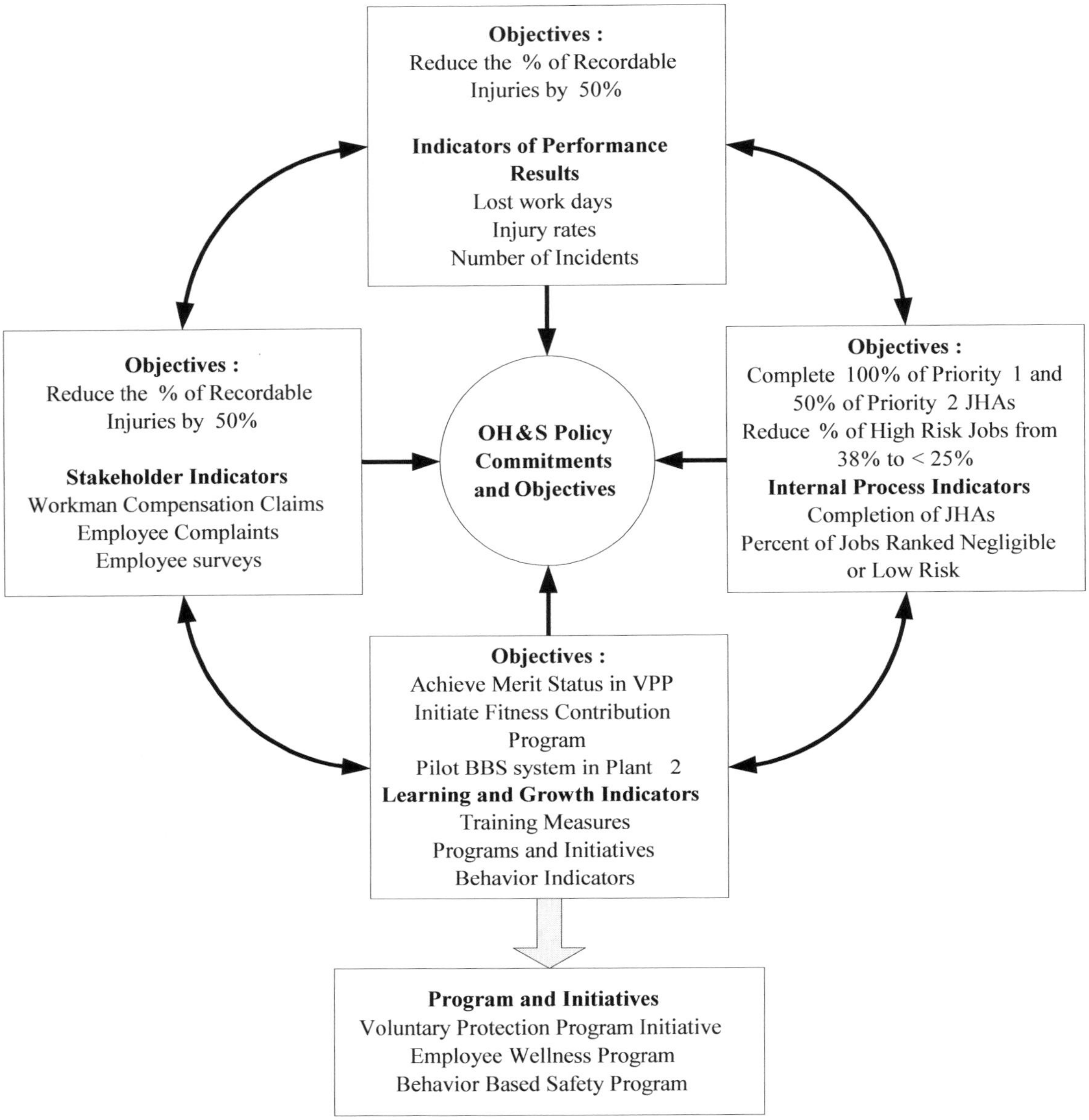

Objectives :
Reduce the % of Recordable
Injuries by 50%

**Indicators of Performance
Results**
Lost work days
Injury rates
Number of Incidents

Objectives :
Reduce the % of Recordable
Injuries by 50%

Stakeholder Indicators
Workman Compensation Claims
Employee Complaints
Employee surveys

**OH & S Policy
Commitments
and Objectives**

Objectives :
Complete 100% of Priority 1 and
50% of Priority 2 JHAs
Reduce % of High Risk Jobs from
38% to < 25%
Internal Process Indicators
Completion of JHAs
Percent of Jobs Ranked Negligible
or Low Risk

Objectives :
Achieve Merit Status in VPP
Initiate Fitness Contribution
Program
Pilot BBS system in Plant 2
Learning and Growth Indicators
Training Measures
Programs and Initiatives
Behavior Indicators

Program and Initiatives
Voluntary Protection Program Initiative
Employee Wellness Program
Behavior Based Safety Program

able to step in and take early action on the drivers of performance. For example, if the management team noted that several departments were falling behind in completing their JHAs and risk assessments or failing to make progress in reducing the percent of jobs deemed high risk, then it could step in and take actions to get performance back on track before too much damage was done.

Note that the examples shown here high level (or strategic) in nature are purely hypothetical. Any number of other objectives and related indicators could be developed. This will be discussed in the next chapter.

Because behaviors are so important to overall safety performance, means of measuring personal behavior should be a part of the organization's overall monitoring. Behaviors drive performance. Some examples of behavior indicators include:

- The extent to which employees are observed complying with or violating safe work practices.
- The extent to which employees participate in health and safety initiatives or improvements.
- The number of suggestions or ideas relating to the OH&S program submitted by employees.
- The extent to which employees watch out for each other and apply peer pressure to follow safe work-place practices.
- The extent to which employees voluntarily report minor injuries and near misses so that these opportunities for improvement can be used.

The challenge with behavior measures is that they can be hard to obtain. Here are the most common methods for obtaining information on these indicators:

- The conduct of daily, weekly, or monthly safety walkthroughs; these walkthroughs can be used to note the number of instances where violations to safe working practices were noted. The common violation will be the failure to wear PPE (e.g., eye protection, hearing protection, etc.), but other violations, such as improper forklift operation, unsafe material handling practices, and unlabeled containers, might also be noted. A checklist is normally used to guide the evaluator in what to look for. The results can be tabulated and a Pareto chart constructed to show both specific areas of concern and long-term trends.
- The number of safety improvements initiated by area safety teams (or 5S teams, Quality Circles, or other standing teams) can be tracked and trended over time.
- The number of suggestions or ideas submitted by employees can be tracked. Note that this metric should be normalized (e.g., average number of suggestions per employee) to account for workforce fluctuations.
- The number of reported near misses can be tracked, although it can be difficult to interpret, because a larger number may be result from a willingness to self-report (a good thing) or lowering of safety standards throughout the plant (a bad thing). This metric should only be used in conjunction with other indicators. For example a rise in the number of reported near misses along with strong evidence of reductions in workplace hazards and associated risks can signal the development of a willingness to self-report.
- Employee safety surveys. Some behaviors, such as employees' willingness in looking out for their co-worker's safety, are almost impossible to measure using common methods. An employee survey, which includes questions such as "To what extent do employees look out for the safety of their coworkers," is one of the best methods to use to gauge such behaviors.

At this point the CFT, the management appointee, and other members of the senior management team must agree on a suite of metrics that will be used to measure the effectiveness and performance of the occupational HSMS. Some of these metrics will be high-level, strategic indicators, such as those shown with the balanced scorecard. Others will be tactical measures used to monitor area performance. Once the metrics to be used have been determined, the systems to collect, analyze, and report the measures must be put in place. A table of safety measures can be developed to summarize the system of metrics used. An example is shown in figure 6.4 and is included in the OH&S planning workbook.

Providing Initial Awareness Training

At this point in the project, some initial awareness training may have already been provided. That session would have focused on the decision to implement an OHSAS 18001 HSMS, including a discussion on how such a system could benefit employees and the organization. If the organization had already developed a policy statement this would also have been discussed. Not known at that time, however, and therefore not yet

Figure 6.4   Occupational Health and Safety Metrics

Summary of Organizational OH&S Metrics					
Metric	Champion	Method	Frequency of Collection	Frequency of Reporting	Reported When
Strategic Metrics					
Lost Workdays	Management Appointee	Summary for entire organization with Pareto by department and type of injury.	Monthly	Monthly and Semiannually	Mo. during Safety Committee Meetings, Semiannually during Management Review
No. Recordable Injuries	Management Appointee	Summary for entire organization with Pareto by department and type of injury.	Monthly	Monthly and Semiannually	Mo. during Safety Committee Meetings, Semiannually during Management Review
No. Workman Compensation Claims	HR Manager	Total number of claims	Quarterly	Semiannually	Management Review
No. Complaints to OSHA	Safety Manager	Summary for entire organization with Pareto by department.	Monthly	Monthly and Semiannually	Mo. during Safety Committee Meetings, Semiannually during Management Review
Employee Safety Survey Results	Management Appointee	Ranked results of annual safety survey.	Monthly	Monthly and Semiannually	Mo. during Safety Committee Meetings, Semiannually during Management Review
Percent JHA Completion	Management Appointee	Summary for organization with breakdown by department	Monthly	Monthly and Semiannually	Mo. during Safety Committee Meetings, Semiannually during Management Review
Percent Jobs Ranked Negligible to Low Risk	Management Appointee	Summary for organization with breakdown by department	Monthly	Monthly and Semiannually	Mo. during Safety Committee Meetings, Semiannually during Management Review
Percent Competencies Achieved	HR Manager	Summary for organization with breakdown by department	Quarterly	Semiannually	Management Review
Performance in Achieving OH&S Objectives	Management Appointee	Measured by percent milestones achieved and overdue	Quarterly	Semiannually	Management Review
Operational Metrics					
Number and Trend of Safety Violations	Safety Manager	Results of weekly safety walkthroughs	Weekly	Monthly	Safety Committee Meeting
Percent Employee Participation	HR Manager	Tabulation of employees involved in safety initiatives	Monthly	Monthly	Safety Committee Meeting
Avg. Number of Safety Suggestions per Employee	HR Manager	Tabulation of employee safety suggestions submitted	Monthly	Monthly	Safety Committee Meeting
Number of Near Misses Reported	Safety Manager	Summary and trend of number of near misses reported	Monthly	Monthly	Safety Committee Meeting
No. Days without a Lost Workday Accident	HR Manager	By department	Monthly	Monthly	Safety Committee Meeting
Results of Internal Audits	Safety Manager	Each audit	Each Audit	Each Audit	Results distributed after each audit, reported during monthly safety committee meeting
Results of OSHA Inspections	Safety Manager	Each OSHA inspection	Each Visit	Each Visit	Results distributed after each visit reported during monthly safety committee meeting
Results of Internal Compliance Reviews	Safety Manager	Each compliance review	Each Review	Each Review	Results distributed after each review reported during monthly safety committee meeting

communicated would be the results of any JHA. Everyone should be made aware of some of the new workplace hazards identified as a result of efforts thus far and the operational controls used to reduce the risks of these activities. This initial awareness training session will focus on communicating this information along with each employee's responsibilities for following procedures and controls listed on the JHA forms.

Topics that should be discussed include the following:

- The organization's OH&S policy statement.
- Examples of major workplace hazards and how each employee can use the JHA to help reduce their risk or exposure to these hazards. While some workplace hazards affect only a few employees and may not warrant discussion at an all-hands briefing, more global hazards, such as slips and falls, fires, and exposure to flying chips or high noise levels, should be briefly discussed so that each employee can see his or her role in performance improvement and personal benefits of compliance. Consideration should be given to citing some statistics here related to the consequences of these hazards to reinforce it.
- The controls (procedures and instructions) that have been developed to control the significant hazards discussed previously. Each employee should be aware of the organization's policy on wearing PPE, emergency response procedures, safe material handling practices, and use of JHAs to detail the hazards in the workplace and controls used to minimize their risks.
- The systems that have been established to allow communication and involvement of the workforce in performance improvement. As noted previously, sustained performance improvement will not occur without the participation of the organization's rank-and-file employees. They need to be instructed as to how they can become involved and participate.

This orientation will typically run between thirty minutes and one hour. It should be reinforced by articles in the company newsletter, staff meetings, and postings on area communication boards as the system is deployed and matured. Lack of such communication will seriously impact the effectiveness of the management system.

Providing Activity-Specific Training

Although the orientation training provided a brief discussion on some of the organization's significant workplace hazards, it did not go into any detail on either the hazards or their associated controls. Because each area will have unique workplace hazards associated with its activities, there will be a need for activity-specific orientation and training. This is best carried out by the area supervisor or team leader. JHA forms, where posted, can assist the supervisor in presenting this orientation. Also included in this training would be any regulatory required training that was identified during the review of legal requirements but which has not been completed.

Remember also that the specification requires competency in tasks that impact on health and safety so detailed training on procedures or systems should also be conducted, where needed, to ensure everyone can safely do their job. This will typically involve training on the procedures, instructions, and the JHAs specific to that area. This activity-specific training should be recorded and shown on the training requirements matrix or other system used for tracking competency attainment.

Conducting the First Round of Internal Management System Audits

With operational controls in place, workforce training completed, and monitoring initiated, the core HSMS is now in place and the first round of internal management system audits should be performed. Plan on conducting at least two rounds of internal audits before any certification assessments. The first round will verify proper implementation, training, and general awareness of the new policies and procedures. Expect to find quite a few weaknesses. Despite best efforts, pockets will be found where individuals or groups will be slow to implement the new controls and practices. This may result from incomplete implementation (e.g., procedures are not available in these areas) or normal resistance to change. Identified weaknesses will be documented so that they can be addressed.

If not already completed, internal auditor training should be provided for the organization's internal auditors. It is often best to schedule this training just before the first round of audits. Auditing is a skill, and skills will be forgotten unless they are applied. This initial audit needs to be a complete system audit and should be accomplished in a short period of time to permit timely reaction to implementation issues. The first round of audits should take no longer than one week and could be accomplished in one or two days if multiple audit teams are used.

Actions resulting from the first complete system audit should be aggressively tracked by the management appointee or audit program manager. There is often a natural let down after passing a major milestone. Some employees may let their guard down and slip into old habits. Do not let this happen. This is the most critical stage in the life cycle of the new program. Fix any weaknesses in the program and continue the push to get results. The greatest gains in performance improvement are often accomplished during the first six months to a year after implementation as there will be a lot of low hanging fruit to pick. Take advantage of the momentum that such victories can provide to solidify the management system.

A second round of audits should be conducted after the corrective actions from the first audit have been taken to verify the effectiveness of the actions and to ensure readiness for the certification audit.

Summary

This relatively short chapter focused on the initial deployment of a HSMS. It was short, because most of the real effort occurred during the planning and design phases. Remember that deployment involves more than simply issuing procedures. It involves verification, training, initial monitoring, auditing, and communication. Communication also involves visible management support for the system by actions and not just words. Leaders who take the time to talk with employees, ask the right questions, demand accountability, and personally demonstrate the right behaviors accomplish more that any set of policies in creating the culture for the HSMS to succeed. This leadership will be especially critical during the early stages of the system deployment, pending success at getting everyone involved in performance improvement.

Part 3 focuses on steps needed to improve the OH&S management system and the organization's health and safety performance, including chapter 7, which describes actions needed to address OHSAS 18001's requirements for objectives and management reviews. Chapter 8 will discuss internal audit strategies, and chapter 9 will present ideas and examples for moving beyond compliance, including OSHA's VPP.

Part 3

OCCUPATIONAL HEALTH AND SAFETY MANAGEMENT SYSTEM IMPROVEMENT

Part 3 focuses on steps needed to improve the HSMS and an organization's OH&S performance, including chapter 7, which describes actions needed to address OHSAS 18001's requirements for setting health and safety objectives and targets and performing management reviews. Chapter 8 will discuss the internal audit process in more detail, and chapter 9 presents ideas for moving beyond compliance using OHSA's VPP.

CHAPTER 7

The Improvement Phase

The improvement phase completes the implementation of an organization's OHSAS 18001 occupational HSMS. During this phase the organization will use the results of its internal audits and performance metrics to assess how well the system is working and where improvement objectives are warranted.

The improvement phase will also complete the first plan-do-study-act circle of improvement as indicated by the inner ring of actions shown in figure 7.1. It also initiates the second round of improvements based on the results of the management review and setting of OH&S objectives as shown by the solid dark arrow and outer ring of figure 7.1. The process will continue as the organization continually looks for ways to improve both its management system and its health and safety performance.

Preparation for the Management Review

The management review is a major event in the life cycle of an organization's HSMS. This may be the only time senior management will be available for a focused review of both the management system and the organization's health and safety performance. Obviously careful planning for this meeting is essential to ensure the greatest benefits are obtained.

There are several considerations relating to scheduling of the management review. First, the review should be scheduled during a day when most, if not all, of the senior management team can attend. As a minimum, the management appointee, the president (or plant manager, if a manufacturing facility), the senior production executive, the chief engineer, human resources manager (or vice president), and the facilities or maintenance manager should be present, along with the safety manager, if this is a dedicated function. If any of these key

Figure 7.1 The Plan-Do-Study-Act Spiral of Improvement

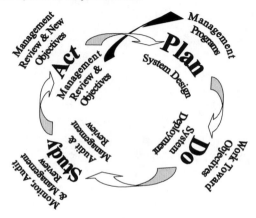

executives are not available then rescheduling should be considered. As many other top managers or executives as are possible should also be available, although it is probably acceptable for them to send alternates if they have schedule conflicts and cannot attend.

Second, plan on spending at least three hours, and possibly longer, to conduct the review. For companies that hold monthly, quarterly, or semiannual management reviews the review might be accomplished in two or three hours. Many companies hold annual reviews, and in this case it will take some time to review and discuss all of the information required. Even when the organization holds several reviews during the year, it is best to have a comprehensive end-of-year review to ensure long-term trends and patterns can be analyzed. It also serves as a great platform for evaluating performance in achieving annual HSMS objectives and to set new objectives for the coming year.

Third, schedule the review to occur shortly before the organization's strategic planning and budgeting process. One of the goals of the management review should be to identify improvement objectives for the coming year. It does not do any good to identify meaningful objectives if there is no money left to fund them, because the budget has already been set.

Once the schedule has been set, it is time to develop the agenda. It is strongly recommended that a formal agenda be created and communicated well in advance of the review. The agenda should list not only the topics to be discussed but also who will lead the discussion of each topic and how much time each topic will be allocated. During the meeting the agenda should be used to focus the meeting and to keep it on track.

A listing of topics that should be discussed during the review includes:

- follow-up from previous reviews,
- internal audit results and the results of reviews of compliance to regulatory and other requirements,
- external communications (including the results of any OSHA inspections or investigations),
- health and safety performance (monitoring and measurement results),
- performance in achieving HSMS objectives, along with any needed adjustments to the objectives,
- status of corrective and preventive actions,
- changes in regulatory or other requirements,
- needed changes to the health and safety policy statement, and
- any other changes to the HSMS needed to meet the organization's commitment to continual improvement.

When discussed in full, these topics should provide ample information to evaluate the management system's overall suitability, adequacy, and effectiveness. The purpose of the review, however, is not to simply note the status of the management system but is to prompt action. The management review provides a unique forum, when those having the responsibility and authority to commit to action are all present and focused on the organization's health and safety performance. A format for a management review agenda is provided in figure 7.2.

Once the agenda has been set the management appointee should begin gathering and summarizing the data that will be presented during the review. To cover the topics on the agenda in the allocated time (about four hours in the example) the presentations will need to be organized. Not every topic will be discussed in detail. For example, the discussion on audit results should not examine every finding from every audit performed during the previous twelve months but should instead focus on trends and patterns in the findings, repeat findings, and those that had special significance. The same could be said for the review of corrective and preventive action, with the primary focus being on any that remain outstanding and on indicators that the actions were effective in eliminating the problems. Summaries are best presented using graphs, charts, and tables. The management appointee should assemble these before the review.

The management appointee will also need to prepare or oversee the preparation of the performance monitoring data for presentation. Although many of these metrics may have been reviewed monthly, this

Figure 7.2   Management Review Agenda

Semi-Annual Management Review Agenda and Report Form

Date of Review 3/02/06
Time: 11:00 AM–3:00 PM
A management review of the OHSAS 18001 management system will be conducted on the date and time indicated above. The review will include the items checked below. Please come prepared to discuss the checked items that you are assigned as champion. Please contact the Management Appointee if you are unable to attend.

Review Items:	Champion	Time
☐ Follow-up actions from previous management reviews	Mgmt. Appointee	20 min.
☐ Results of system audits (internal and external)	Audit Program Manager	20 min.
☐ Results of Compliance Reviews	Mgmt. Appointee	20 min.
☐ Summary of communications and feedback		20 min
☐ From external parties	Mgmt. Appointee	10 min.
☐ From employees	HR Manager	10 min.
☐ Health and Safety process performance including trends		30 min.
☐ Lost Workdays and Recordable Injuries	Facilities Manager	
☐ Workman Compensation Claims	HR Manager	
☐ OSHA Complaints	Safety Manager	
☐ Safety Survey Results	Mgmt. Appointee	
☐ JHA Completion and Risk Reduction	Mgmt. Appointee	
☐ Training Competencies	HR Manager	
☐ Status of significant preventive and corrective actions	Safety Manager	15 min.
☐ Status of any new or pending regulations	Safety Manager	10 min.
☐ Changes that could affect the OH&S system	All	5 min.
☐ Status in achieving OH&S objectives	Mgmt. Appointee	30 min.
☐ Recommendations for improvement	All	15 min.
☐ New OH&S objectives	All	30 min.
☐ Other (List)	All	30 min.

Summary of Actions and Agreements Reached:

Action	Who	By When	Date Complete

Note: Include any carryover items from the previous review that were not completed (assign new estimated completion date).

Distribution:

President	Human Resources Manager
Management Appointee	Safety Manager
VP Personnel	Production Manager
VP Production	Engineering Manager
CFO	Audit Program Manager
COO	Facilities Manager

review should focus on long-term trends and comparisons with previous periods. The management appointee should develop a specification format for data presentation that shows current performance, performance trends over the past twelve months, and perhaps annual comparisons over the last three years. Once the format has been prepared it can be provided to others responsible for presenting their data during the review. An example of one format is shown in figure 7.3.

The format shown in figure 7.3 allows the management team to quickly assess current performance and trends in completing JHAs. Notice that the graph also shows the goals for JHA completion. These are aligned with the organization's HSMS objectives, which targeted 100 percent of the priority-one JHAs and 50 percent of the priority-two JHAs to be completed by the end of the year. The management team can easily see that performance in completing priority-one JHAs is ahead of schedule, whereas performance in completing the priority-two JHAs is just starting to slip. This gives them the opportunity to dig deeper into the reasons for the stalled performance before it falls too far behind. The report could be generated automatically using information from the job lists in the OH&S planning workbook using a Microsoft Excel macro. Note also that the number of JHAs changes over time as new jobs are identified and old jobs eliminated.

To prepare for the discussion on pending regulatory changes that could impact the system the manage-

Figure 7.3   Format for Displaying Performance Data

JHA Completion	Jan	Feb	Mar	Apr	May	Jun	Jul	Aug	Sep	Oct	Nov	Dec
Total Priority 1	74	74	76	76	76	79						
Complete Priority 1	8	23	32	39	49	57						
Pri. 1 Percent Complete	0.108	0.311	0.421	0.513	0.645	0.722						
Total Priority 2	132	134	135	135	135	135						
Complete Priority 2	12	12	22	25	28	31						
Pri. 2 Percent Complete	0.091	0.09	0.163	0.185	0.207	0.23						
Total Jobs	302	304	311	308	310	310						
No. Complete	20	35	51	62	68	84						
All Jobs Percent Complete	0.066	0.115	0.164	0.201	0.219	0.271						

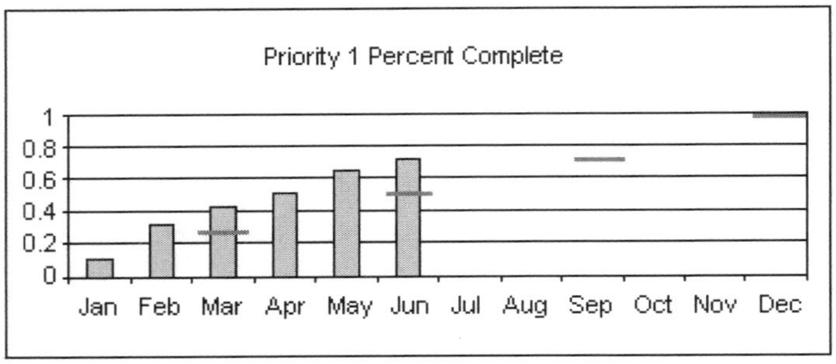

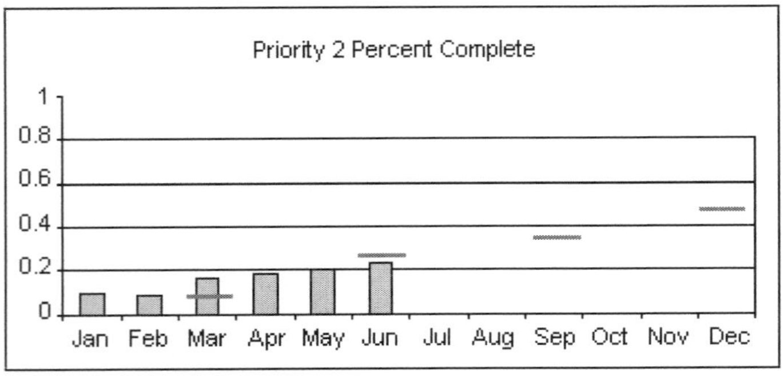

Priority 1 Percent Complete

Priority 2 Percent Complete

ment representative should talk with the company's attorney (if it has one), outside consultant, or he or she can visit the federal or state websites where new and upcoming regulatory changes are normally posted. Planned facility modifications, expansions, new products, and process lines should be considered to help plan for the discussion on changes that could impact the HSMS.

The summary of internal and external communications should be sourced from the organization's communication log or other system for collecting the views of outside interested parties and employees. Complaints, concerns, and suggestions for improvement should all be considered during the review.

Once the material for the review has been assembled the management appointee should consider providing a copy of the presentation and its charts, graphs, and summaries to attendees a week or so before the review if possible. This will give the management team a chance to scan the information and identify areas of concerns or questions before conducting the review. This will lead to a more focused and timely management review. This preview of the data cannot be allowed to become a surrogate for the actual review however. If the management appointee has indications or experience that some senior managers will not attend the review if given a chance to see the data beforehand, then do not distribute the data before the review. Remember that the purpose of the review is to promote discussion and action, not to simply conduct a status review. This cannot happen unless the senior management team actually meets. Always provide a copy of the agenda at least two and preferably four weeks before the review to each attendee. This allows them time to prepare for their assigned areas and to clear their schedule.

Conducting the Management Review

The agenda should serve as the template for conducting the actual review. Follow the agenda. Use basic meeting management skills to keep the review focused and on track. Because everyone has received the agenda, knows what topics they are responsible for, and how much time they have to summarize that topic, the meeting should flow smoothly. Allow additional time for serious discussions as long as they are focused on actions. Note that there is an additional thirty minutes for other items at the end of the example agenda. Always build in some extra time to allow for discussion as needed.

Check off each item as it is discussed. You will also need to generate meeting minutes or records of the review. These records should not record the details of each topic discussed but rather should summarize the overall discussion. Together with the presentation package (graphs, charts, etc.) used for the review and the actions matrix on the form, this summary will provide adequate documentation of what was discussed, the results of the discussion, and the details of any actions, decisions, or agreements reached. I prefer to use an electronic copy of the agenda to keep my meeting minutes. A short summary of each topic can be inserted underneath each topic. This could be done by hand by inserting five or six spaces between each topic in the form and then printing it out before the meeting for use in taking notes or someone may be assigned to enter the notes directly with a laptop during the meeting.

Especially important is that any and all actions, decisions, and agreements reached during the management review be fully documented. The example form provided has a table inserted to accomplish this. The management appointee should review each action item at the end of the meeting before disbanding the team to ensure there is no misunderstanding or miscommunication as to who will do what by when. Again, the purpose of the review is to drive action. Make sure everyone agrees on these actions and who will be responsible for them.

You may have noticed that we have not discussed the evaluation of performance in achieving objectives or the identification of new objectives. These absolutely should be a part of the management review, but the importance of the topic warrants a separate section.

Health and Safety Objectives and Management Programs

We are now ready to put the last link into the circular chain that is the HSMS. Keep in mind that the actions discussed in this section are in fact happening during the management review. Although it is possible to have a separate review for objectives and management programs, the management review provides an excellent time to evaluate and set new objectives, because the performance of the overall management system has just been analyzed, discussed, and opportunities to improve it identified. With this background, the identification of new objectives should be relatively straightforward.

For a new 18001 management system no objectives will have been set so a review of performance in achieving the objectives will not be needed. The discussion that follows focuses therefore on the setting of initial objectives and the development of management programs to ensure their achievement.

When setting the objectives keep in mind the requirements of the specification and the discussion provided in chapter 6 regarding the identification of a set of balanced indicators or performance in alignment with the organization's policy commitments. As noted in chapter 2, the specification requires that the following items be considered as part of the setting of objectives:

- the commitments the organization made in its policy statement,
- the organization's legal and other requirements,
- the organization's hazards and risk,
- the organization's financial, business, and operational requirements, and
- the views of interested parties.

Each of these items should have been thoroughly covered during the earlier topics of the management review. The challenge now is to identify meaningful objectives that will drive improvements in performance. Now that the organization has several months' worth of data it should be in a position to select meaningful objectives. Several examples of strategic objectives were noted in chapter 6. Other examples of organizational objectives might include:

- reduce the number of slip and fall accidents by 75 percent;
- organize and conduct a seat belt campaign to reduce the injuries to workers and their families from automobile accidents;
- complete a 5S safety campaign in all manufacturing areas;
- certify all production supervisors in basic cardiopulmonary resuscitation;
- implement an ongoing fitness and nutrition program in partnership with a local hospital;
- construct an on-site fitness facility; and
- design, implement, and promote an organizationwide Safety First campaign.

As you can see from this and the previous examples the scope and nature of potential objectives can vary widely. The status of the organization's existing health and safety program should drive the nature of the objectives. If the organization is currently wrestling with a higher than expected number of accidents or incidents then objectives focused on JHA completion or risk reduction and focused improvements in targeted areas, such as slips and falls, are more appropriate. If the organization has its accident rate well below industry averages and is well along in its risk reduction efforts, then more focus can be directed at long-term employee health and fitness programs. Likewise the number of objectives identified can vary. Some organizations attempt to set at least one objective for each policy commitment each cycle. Others target a few major projects and a number of smaller objectives. Notice that the 18001 specification does not provide a minimum number. At a minimum an organization must have at least one; if only one, it had better be a big one.

The specification also requires that the objectives be set at each relevant level and function within the organization. Some of the objectives in the bulleted list above would affect the entire organization and, therefore, apply to each function and all levels. Others are more focused, such as the 5S safety campaign, which may only affect the production function. The key is to communicate the objectives and to develop plans or targets at each affected function and level.

With these objectives identified the champions can now develop detailed action plans for achieving the objectives. These plans should include the objective, related actions that need to be accomplished to achieve the objective, responsibilities for these actions, and timing for completion of actions. When complete, this program of objectives, actions, time frames, and responsibilities form management programs suitable for planning and monitoring the actions needed to achieve the overall objective.

These management programs could take the form of a Microsoft project plan or an Excel spreadsheet. There is no specified format that must be used for the management program. An example of a simple format is provided in the health and safety planning workbook provided on the CD and is shown in figure 7.4. The important thing to keep in mind is that the management program must provide sufficient detail to allow monitoring of performance in achieving the objective. This means that several intermediate milestone actions should be identified.

Normally the details of the management program including actions, responsibilities, and intermediate milestone dates would be developed after the management review was completed. The goal of the management review is to simply agree with the senior management team on what the overall objectives should be for the coming period. Once the management programs have been developed, the objectives would be funded and monitored during other periodic reviews. The importance of scheduling the management review and the setting of objectives before the annual strategic planning and budgeting process should now be obvious, as several

Figure 7.4   Health and Safety Management Program Using Excel

Environmental Management Programs   Date Revised: 6/13/01

Objective	Major Phases	Action	Responsibility	Due Date	Comp. Date	Overdue?	Comments
Champion M. Jacoby							
Certify all production supervisors in basic CPR and first aid by September, 2006	Organization	Identify the components of the program	M. Jacoby	1/15/2007	1/10/2006	NO	
		Establish cost and budget	M. Jacoby	1/30/2007	1/20/2006	NO	
		Generate list of participants	M. Jacoby, C. Sheehan	1/30/2006	2/15/2006	NO	
		Identify training providers	M. Jacoby	2/28/2006	3/10/2006	NO	
		Select training provider	T. Tomlines	3/30/2006		YES	
	Program Development	Contract with trainer	B. Smith	4/15/2006		NO	
		Verify program content	B. Smith	4/30/2006		NO	
		Develop program content	Contract Trainer	5/20/2006		NO	
		Approve final course	C, Sheehan	5/30/2006		NO	
		Schedule classes	B. Smith	5/30/2006		NO	
	Training and Certification	Complete basic CPR training	M. Jacoby	6/30/2006		NO	
		Complete basic first aid training	M. Jacoby	8/15/2006		NO	
		Complete bloodborne pathogen training	M. Jacoby	9/15/2006		NO	
		Receive certificates	M. Jacoby	9/20/2006		NO	
		Recognize graduates	M. Jacoby	9/30/2006		NO	
Percent Overdue Percent Not Overdue		Number Overdue		1	Copy and paste the formula in cell J3 and N3 into their corresponding cells of any new rows inserted to add new actions.		
		Number Not Overdue		14			
		Total Actions		15			

of the objectives will require funding to accomplish. With the senior management team's heavy involvement in selecting the objectives, this funding should be easier to obtain.

Summary

We have now completed the design, deployment, and planning for the initial round of improvements of the 18001 management system. You understand how the system is performing and where improvements are needed, and you set objectives and developed corresponding management programs to achieve them. The next chapter will focus on the details of performing OH&S management system audits as the basis for ongoing improvement of the 18001 management system.

CHAPTER 8

Conducting Health and Safety Management System Audits

This chapter will build on the discussion of auditing in previous chapters by presenting strategies for the performance of audits of HSMSs based on the OHSAS 18001:1999 specification.

OHSAS 18001 Revisited

As noted in chapter 2, the OHSAS 18001:1999 specification is based on five primary clauses as shown in figure 8.1.

These five clauses can be grouped into five operational components as follows:

- A planning component. In OHSAS 18001, planning is represented by the ongoing identification of job hazards, the assessment of the risks associated with these hazards, and the identification of appropriate methods to control them. Planning for unexpected events is captured by the emergency preparedness and response requirements. Planning is centered on the establishment of an organizational policy that declares the company's commitments for health and safety performance.
- An operational component. In OHSAS 18001, requirements focus on implementation and maintenance of the controls are specified in the requirements for implementation and operation.
- A monitoring and corrective action component. In OHSAS 18001 these activities are defined in performance monitoring and measurement, nonconformance, and corrective and preventive action, and the internal audit clauses.
- An improvement process. Improvement is embedded in requirements for the setting of health and safety objectives, management programs, and in the conduct of management reviews.
- Key support processes. These activities support the overall HSMS and include document and record control, structure and responsibility, communication, training, awareness, and competence. They are found in the implementation and operation and monitoring and measurement sections of the specification.

In figure 8.2 the organization's core business activities are represented by the dashed lines. They represent the on-the-job activities and facilities that can pose risks to an organization's employees. The OHSAS 18001 components that monitor and control the health and safety risks of these activities are then shown interacting with these core business activities.

Audit Scope

The audit scope defines what is to be included in the audit. It is recommended that an organization groups related OHSAS 18001 components together for audits for two reasons. First, it provides for a more efficient

Figure 8.1   OHSAS 18001 Health and Safety Management System

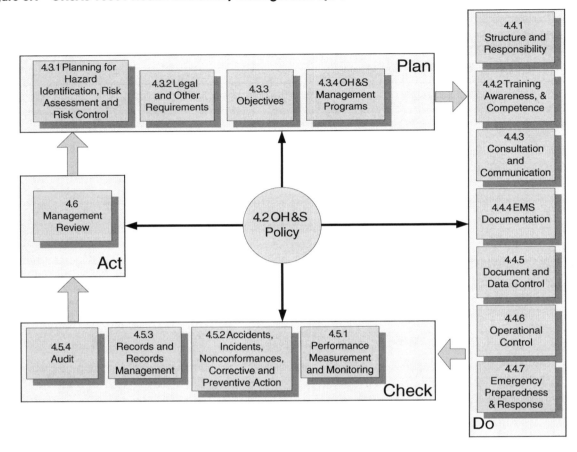

and effective audit. The grouped elements are closely related, and it can be hard to evaluate one without evaluating the others in the group. Second, this supports the application of the process approach popularized by ISO 9001 to HSMS audits. By grouping these elements together, natural processes where the output from one process becomes the input into a related process are formed. The recommended audit groupings are a little different from the operational groupings shown in figure 8.2, as will be discussed.

HEALTH AND SAFETY PLANNING COMPONENTS

These are normally evaluated first, because the proper identification of health and safety hazards and assessment of their risks is central to all other components within the OH&S system. In figure 8.2 inputs into the health and safety planning process include the organization's planning for safe processes, and the safety of the processes, equipment, and facilities themselves. Also key to planning is the identification of legal and other requirements and planning for emergencies. All of these activities are folded around the health and safety policy statement.

The audit strategy differs slightly from figure 8.2 in that an organization should also consider evaluation of compliance to safety regulations during an audit of the planning components. There is a natural audit flow that arises from doing the identification of legal and other requirements and evaluation of compliance together.

Figure 8.2   OHSAS 18001 Process Flow

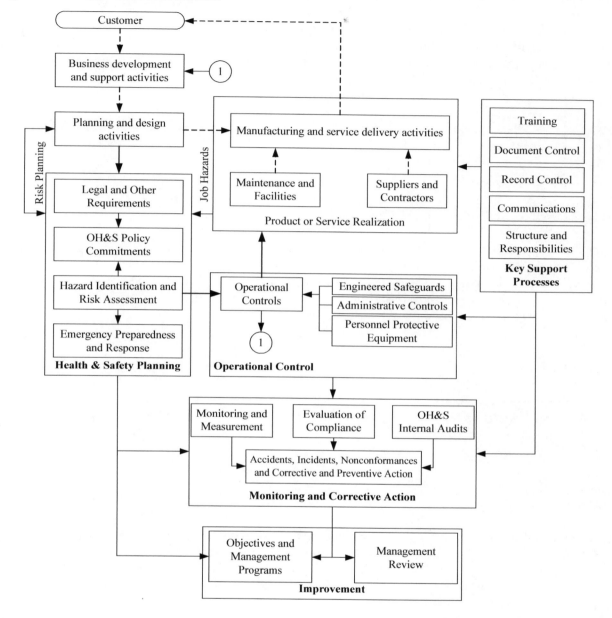

This flow will be presented during the discussion of the evaluation of the health and safety planning components.

OPERATIONAL CONTROL

The operational control component verifies that the controls associated with the organization's significant hazards are in place and are being used and maintained. It is similar to the production audit of a QMS. It answers the question: *"Are we doing what we said we would do?"* In any organization there will be multiple activities that need to be controlled, and there could be multiple controls for each activity. These audits will evaluate these controls.

The audit strategy departs slightly from figure 8.2 in that it also evaluates a portion of the monitoring and measurement component during operational control evaluations, and the auditor may also spot-check compliance to regulatory requirements. If the organization has trained internal auditors who are also knowledgeable of health and safety regulatory requirements, then the health and safety compliance reviews could be folded into these operational control audits *almost* entirely.

Note that during these reviews the auditor will also be checking on the control of documents and records associated with the activities being evaluating, the training and competency of personnel performing these activities, and their knowledge of their roles and responsibilities for properly conducting the activities in a safe manner.

MONITORING AND CORRECTIVE ACTION

This audit evaluates the overall monitoring and performance of the HSMS. It includes an analysis of overall trends in health and safety performance and actions being taken to correct any adverse trends. It also evaluates the internal audit program and the follow-up on any weaknesses noted during these audits. Both of these components feed into the accidents, incidents, and corrective and preventive action system, which will also be reviewed during this evaluation.

IMPROVEMENT

The evaluation of improvement activities will focus on the setting of health and safety objectives and the establishment of health and safety management programs to achieve them. Performance toward meeting these objectives will be assessed along with any actions being taken to ensure they are met if performance is lacking. The management review process will also be examined.

KEY SUPPORT PROCESSES

The centralized activities for key support processes, such as document and record control, will be evaluated in this audit. Only the common or centralized control portions of these systems need be evaluated, because many of the outputs from these activities (accurate documents located where needed, legible and complete records, etc.) will be assessed during the other audits. Other items included in this review will be roles and responsibilities of key functions, communications, and training.

The remainder of this chapter will focus on strategies for conducting audits of these groupings. The audit checklists provided on the CD accompanying book have been structured along these groupings. The reader is encouraged to review these checklists as they read the discussions that follow.

Health and Safety Planning

The planning process forms the heart of the HSMS. The activities that form this process are shown in figure 8.3. This figure also outlines the key audit points that will be examined during the evaluation.

The reader should note the parallel path through the health and safety planning process. The OHSAS

Figure 8.3   Health and Safety Planning Activities

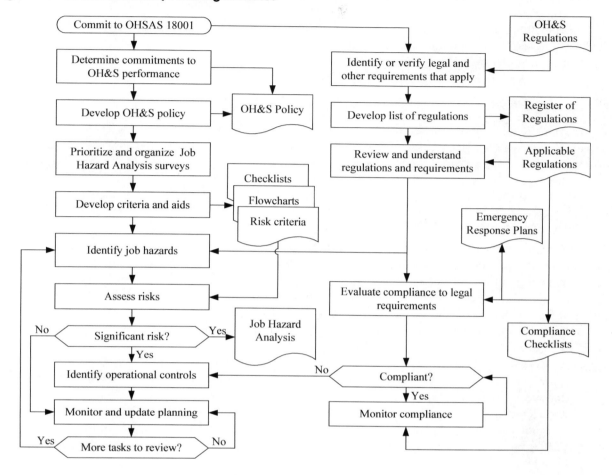

18001 specification requires an organization to meet its regulatory obligations. In the course of implementing OHSAS 18001, a company will have to demonstrate that it understands what health and safety regulations it must meet, has access to what those regulations require, and is performing reviews to ensure it is meeting those requirements. For many small- to medium-sized companies without health and safety specialists on staff, this may be the first time that they have systematically evaluated their health and safety compliance posture. What they learn must be incorporated into the planning for the HSMS. This health and safety compliance review may be done before the identification of job hazards, or it may be performed as part of the initial health and safety review. The key audit point is that it must be done as part of the planning of the OH&S system and any deficiencies corrected or prevented through the application of corrective action and operational controls.

One of the key differences between an occupational HSMS and an EMS lies in the initial identification of risks. Whereas in an EMS the identification of all significant aspects can be accomplished during the initial environmental review, in an OH&S system this identification will be ongoing. The systematic review of all workplace hazards may take many months or even years. In essence it is an ongoing activity because of the constantly changing nature of the workplace. The initial review should identify the jobs, tasks, and areas that must be evaluated and should prioritize them based on their perceived risk. The organization then conducts JHA on these tasks in a prioritized fashion. I do not recommend waiting until all of the JHAs have been completed before implementation of the rest of the management system. To do so would be to lose out on the significant benefits that an 18001 based system can provide during this lengthy time frame.

AUDIT STRATEGY AND KEY AUDIT POINTS

This section will define a general audit flow and strategy that can be used to evaluate the health and safety planning process. The components of this process will be presented in the order in which the auditor would normally evaluate them.

Health and Safety Policy

The auditor should begin the audit by examining the health and safety policy statement. The policy statement, or more important the objectives and commitments within the policy, form the heart of the planning process, which in turn forms the heart of the HSMS. Rather than evaluating simple compliance to the specification, the auditor should determine if the policy and its development meets the intent of the specification. Compliance to the intent of the OHSAS 18001 specification can be evaluated by examining the following key audit points:

- Who was involved in the development of the policy statement? The specification requires top management's involvement in the development of the health and safety policy statement. There is no hard and fast definition of who makes up top management, but a general thumb rule often used is that top management is comprised of the senior person present at a facility and his or her direct reports.

   This involvement should be more than simply signing off on a policy developed by the health and safety management representative or a consultant. It should include active debate about what health and safety objectives the organization wants to achieve and what commitments it is willing to make. Later on, the company will be required to establish improvement objectives focused on meeting its commitments in the policy statement. Most improvement projects require some initial allocation of time, money, and other resources. Management should not make commitments that it is not willing to back up with sufficient resources. The auditor can assess the level of top management involvement by asking senior management how the policy was developed. If management's involvement was simply a signoff and they admit to this, then the auditor should consider issuance of an OFI. Engaging top management in the development of the health and safety policy will help to ensure the alignment and support needed for real health and safety performance improvement.
- Is the policy statement appropriate, given the nature and scale of its and health and safety risks? This is a difficult requirement to evaluate, because it is somewhat subjective. Unless the policy statement makes no sense, most auditors would not challenge it.
- Health and safety policy objectives and commitments. These form the core of the policy statement. They represent what the organization feels is important regarding its health and safety performance. They can also form the framework for health and safety objectives and related management programs.

   Unfortunately, most organizations do little more than embed the two mandatory commitments from the OHSAS 18001 specification in their policy. The two commitments, continual improvement and compliance to legal and other requirements, are broad and make it difficult to provide focus on what is most important to the organization. What an organization wants to do is to use the policy commitments to help focus on what is most important to them.

   The auditor should of course verify that the two mandatory commitments are found somewhere in the policy statement. Ideally, the auditor will also find a few more specific commitments have been added beyond the two mandatory ones to drive performance in certain key areas. Some examples of specific commitments include:
   - We are committed to the long-term safety and health of our employees. This commitment goes beyond basic safety and health and could be used to drive smoking cessation programs, nutrition programs, and fitness programs.
   - We are committed to the development and maintenance of a culture of health and safety through-

out the organization. This commitment could be used to drive the implementation of a behavior-based health and safety system.

○ We are committed to the development of safe products that enhance the quality of life of our consumers. This commitment would be appropriate for a firm that designs the products that it sells. It could be used to drive the design of products that go beyond meeting regulatory safety requirements.

These are but a few of the specific commitments that might be offered up in the health and safety policy statement. Many others were noted in previous chapters. Their usefulness is in their ability to focus health and safety objectives toward specific improvements that have a large and real effect on the organization's health and safety performance. They point to where an organization can really make a difference beyond meeting minimum safety laws.

To summarize, the auditor should verify that the mandatory commitments are embedded within the policy statement. Failure to do so should result in an NC. The auditor should also evaluate how well the policy commitments reflect the organization's attitude toward health and safety. If found lacking, the auditor should issue an OFI.

• Documented, implemented, maintained, and communicated within the organization and to the general public. All policy statements are documented, or at least I have not found one yet that was not. Implementation, maintenance, and communication within the organization, including people working on behalf of the organization, will be evaluated during other audits. This leaves communication to the general public.

The auditor should ask how the policy is made available to the general public. Ask, "If my Aunt Betty, who drives by the plant every day and sees the OHSAS 18001 flag flying, calls and asks about the health and safety policy, what would you tell her?" Answers may be that she would be directed to the company website where it is listed, she would be faxed a copy, or that the call would be directed to the health and safety management appointee who would send her a copy or read it over the phone. The key is there has to be a process to make it available to the public. Almost every company has a website. Posting the policy on the website is one of the easiest and most effective ways of communicating to the public. The auditor should issue an OFI if the process for communicating the policy is cumbersome or inconvenient. For example, if the answer to Aunt Betty's question was "Well, we would tell her to stop by and come into the lobby. It is posted there." Although minimally compliant, it is neither practical nor realistic, and should result in an OFI.

Health and Safety Hazards

If the policy is the heart of the OH&S system, then the identification of significant health and safety hazards is the head. The goal of this process is to identify those activities that must be controlled to prevent harm to the company's employees or the general public.

As noted in the introduction to this chapter, the identification of workplace hazards and determination of their related risks will be an ongoing activity. The auditor should verify that a systematic process is in place to identify and evaluate these hazards and that progress is being made to complete the JHAs throughout the workplace.

The auditor should start by asking to see the procedure for identifying significant health and safety hazards. The specification requires that an organization have a procedure, but it does not necessarily have to be documented (although it is certainly recommended). If the process has been documented, then the auditor should ask the health and safety management representative, management appointee, or process owner to walk him or her through the procedure, stopping at key points as outlined in this section. If it has not been documented, then the auditor should still ask the auditee to walk him or her through the process, paying close attention to the key activities and making notes to follow up on when asking to see the **objective evidence**. The auditor will have to rely more on the specification and on his or her checklist in this situation.

During the walkthrough, verify the following key elements of the process:

- Is the strategy for identifying workplace hazards comprehensive? The auditor should verify that all areas, core activities, and processes are included in the evaluation. Many organizations conduct the review by portioning the plant into areas, such as production areas, administrative areas, outside facilities and grounds, and auxiliary areas, such as tank farms, pumps, and boilers. Others structure the review around job functions, such as warehousemen, production operators, maintenance, administrative, management, and clerical functions. The auditor can look for evidence that each of these areas or functions were included in the review or in plans for future reviews. The auditor should inquire about any areas where there is no evidence of inclusion, either in the JHAs already performed or in the plans for future reviews. If it becomes clear that major areas or functions were not included in the review, generate a finding. An example of a job listing that identifies the activities to be reviewed for an area, along with priorities for completing them, was discussed in chapter 4 and is included in the OH&S planning workbook. In this system, the strategy would be to first evaluate and analyze all priority-one tasks, followed by priority two and then priority three. The red (R), yellow (Y), and green (G) column provides information on the status of completing the reviews. Criteria for assignment of R, Y, or G are contained in a comment on the spreadsheet. Using this and the spreadsheets for the other areas of the facility, the auditor could establish whether a comprehensive system was in place to evaluate workplace hazards.

  Another consideration is whether the evaluation included nonroutine as well as routine activities. Nonroutine activities could include bulk liquid loading and unloading, maintenance of equipment and facilities (whether performed by the organization or by contractors), and facility modification, expansion, or renovation.

- Were the reviews thorough? Once the auditor is satisfied that the scope of the review was adequate he or she may wish to verify that the team that performed the review knew what to look for. The auditor may ask what training the team received before conducting the review. Personnel performing JHA should have training before conducting the evaluations. He or she may also ask to see any checklists used to guide the analysts in what to look for.

  Related to this topic is evaluation of who was involved in the review. Ideally, the JHA will be performed by CFTs with representatives from each of the areas evaluated. Many companies train and then use safety committee members or supervisors to perform the JHAs. This can be especially important to identifying the nonroutine activities that should be considered but which may not be going on when the analysis is performed. If the auditor discovers that the review was performed by one person, then he or she should more aggressively evaluate the comprehensive and thoroughness of the review by examining the evidence already cited. Nothing in the specification *requires* a CFT be used to conduct the review, but it is highly unlikely that one person can do the job adequately. Because it is possible that one individual is the source of all knowledge, the auditor must focus on the results.

  Completed JHAs could be sampled and reviewed by the auditor for an indication of the thoroughness of the analysis. Other methodologies, such as HAZOP or FMEA, are also sometimes encountered as methods to determine hazards and their related risks.

  If the auditor finds evidence that the JHAs failed to consider important hazards, or that the reviews did not include obvious hazards, like chemicals, ergonomics, fire, and explosion hazards, then the auditor should consider issuance of a finding. Note also that the OHSAS 18001 specification requires employee involvement in the development and review of policies and procedures used to manage risk. Using employees representing all areas of the organization is another way to demonstrate involvement of employees in the design of the OH&S management system, although it is not required.

- How were risks determined and classified? The auditor should examine the rationale used to determine and classify the risks associated with the hazards identified during the JHA. The most common method for assessing risks combines the likelihood of occurrence of the hazard with its severity if it does occur. The results of this analysis are then used to classify risks as negligible, tolerable, moderate, high, and unacceptable. Actions and controls to minimize risks can then be focused on unacceptable, high, and then moderately risky activities in that order. Tolerable risks can be lowered as time and resources permit. This is by no means the only way to assess risk, but the organization must have some systematic method that provides a reasonable and actionable assignment of risk categories to be in compliance with the specification and to determine where actions need to be taken.

  The key audit point is that the methods used to determine what is and what is not a significant risk should be reproducible. There should be a consistent process in place. A team evaluating the same health and safety hazards next year should come to roughly the same conclusions as to what is a significant risk and must be controlled. Consistency is important, because the jobs change often within a dynamic system. Processes, products, and facilities change. Health and safety concerns change. A consistent set of criteria helps to provide the basis for these changes. Verify that there is some method to the assignment of risk. If there was no rationale at all, generate an NC, citing the lack of a procedure (i.e., process). If there was some reasonable rationale, but it was not formalized or applied consistently, consider generation of an OFI.

- How is information kept up to date? For audits that are performed well after the OH&S system has been in place and certified, this question becomes a central focus of the evaluation. Up to this point, the audit flow has focused on initial JHA. For an existing system, the auditor should have confidence that the methods used to identify health and safety hazards and their related risks have been verified to be adequate. The main question for an existing system, therefore, is whether these methods are being applied to product, process, and facility changes. The key questions are how are those responsible for reviewing new products, services, and activities made aware of the need to reperform a JHA.

  The auditor should begin by first identifying new or modified products, services, or activities and facility modifications that have taken place since the last audit. If the auditor does not know what they were, then he or she should first talk to the production manager, maintenance manager, or engineering manager to help generate a list of recent modifications, job redesigns, or new product or process introductions. These items will become the auditor's samples for later verification during the interview.

  The auditor should then interview the management representative or management appointee about the methods used to keep the JHA current. Many organizations use a "management-of-change" process to control and plan for new products, processes, and products. This process could be modified to include consideration of job changes that could impact the health and safety of employees. Others have an advanced product quality planning process. Some organizations use a management of change and an annual review to ensure new hazards are identified. The auditor should thoroughly understand the methods used by asking questions and by asking to see the documents, checklists, and other aids used to control changes. The absences of any consistent method to stay on top of new hazards should result in a finding.

  Once the auditor fully understands the methods used to stay up to date, then he or she should test the application of the methods by asking to see the results of the reviews for the items selected before the audit. If a management-of-change process was used, ask to see the checklist completed for the facility or process modification made two months ago. Assuming that new health and safety hazards were identified for the change and then the auditor should ask to see the results of their risk analysis. If the auditor finds that the process was not followed, or that the management representative was not even aware of significant changes to the company's tasks, services, activities, or facilities when they should have been, issue an NC.

  Staying aware of changes to processes that could impact on the OH&S system is important from a regulatory standpoint as well as a health and safety performance perspective. The organization is required by law to maintain a healthy and safe workplace. Recognizing new hazards and putting in place controls to minimize their risks is a regulatory requirement.

The bottom line is that the auditor should verify that methods are in place and are being used to keep their organization's JHAs current.

- Operational controls. The auditor also needs to verify that the organization has taken its significant health and safety risks into account in the design, implementation, and operation of its OH&S system. This can be evidenced by the identification and implementation of appropriate controls to minimize or lower the risk associated with its activities and facilities. The auditor can ask how significant health and safety hazards are controlled. At this stage, the auditor is most concerned that appropriate controls have been identified. The operational control audits will confirm that they were properly implemented and are being maintained.

Legal and Other Requirements, Evaluation of Compliance

One of the first areas where organizations see a significant improvement after implementing OHSAS 18001 is in their health and safety compliance posture. This is especially true in small- to medium-sized firms that do not have health and safety professionals on staff. In these organizations the identification of the legal requirements that apply to their activities may require more effort than the implementation of the HSMS itself. Even so, it is critically important that the organization not only identify the regulatory requirements it must meet but also understand them and ensure that they meet them. This portion of the health and safety planning audit focuses on verification that the organization is meeting its policy commitment to comply with legal and other requirements.

Other requirements in this sense mean requirements imposed by the management system design itself, by customers, and by voluntary adoption of health and safety protocols such as OSHA's VPP. Internal requirements will be called out in the organization's policies and procedures. Customer requirements will be called out in contracts and specifications. Voluntary protocols, like the VPP should be listed along with the legal requirements that the organization must meet.

The auditor should begin by asking the management representative, management appointee, or individual responsible for health and safety compliance for a listing of the regulations and other requirements that apply to the organization. While the specification does not require a listing, or register of regulations as it is often called, typically there are far too many to commit to memory. As a result, a listing of some type is a practical necessity. Once the auditor has this list, verify conformance to the specification by evaluating the following:

- Is the listing comprehensive and complete? The auditor should verify that the listing includes federal, state, and local requirements. Most states have the authority to implement the federal requirements and do so by enacting state statutes that must be followed. If the state has the authority to implement the federal regulations, then listing the state statutes and public acts may be sufficient, because they must be at least as restrictive and are sometime more stringent than are the federal regulations. Local requirements may include local building codes and fire and evacuation regulations. As noted, I also look to see if significant voluntary requirements, like the VPP are listed, assuming they apply.

- Is the listing sufficiently detailed? Simply listing the Occupational Safety and Health Act does not provide evidence that the organization understands the legal requirements it must meet. Ask the auditee "Which parts of the Act apply to us?" Some organizations simply list the name of the law or CFR that contains the requirements. The Occupational Safety and Health Act is hundreds of pages long and may contain only a few sections that are applicable to the organization. Some of the many provisions of the Occupational Safety and Health Act were listed at the end of chapter 3. Not all of these will apply to every organization.

The auditor can evaluate whether the auditee has sufficiently researched the regulations by asking the auditee to show him or her the requirements or standards that apply. Pick three or four regulations from the list to sample. This question evaluates two critical prerequisites of health and safety compliance—does the organization have access to the regulations it must meet (also a specification requirement) and does the organization understand what they require? Often when the auditor asks this question they get a blank stare or the auditee pulls up 29 CFR 1900 through 1910 and tries to find his or her way through the regulations for the first time. An organization does not stand much of a chance of meeting the requirements if they do not know what they say. Of course, the auditor must ensure that he or she is talking to the right person. If the auditee referred you to another employee and that employee had the knowledge and was involved in ensuring compliance then that would be acceptable.

Many organizations now use hyperlinks to the federal or state website where the regulations are maintained. This is good practice, because it is practical and greatly simplifies staying up to date with regulatory changes. Local regulations and laws normally have to be retained in hard-copy format. If the organization meets the access requirements of OHSAS 18001 by keeping hard copies of the CFRs, then verify that these documents are controlled and maintained up to date. One of the challenges of using hard copies of the CFRs is keeping them current. They are revised on an annual basis and can be expensive.

- Is there evidence that the organization is evaluating its compliance to regulatory and other requirements? Other requirements, such as VPP, are normally evaluated during other evaluations. Compliance to regulatory requirements is typically verified through the use of health and safety compliance reviews. Although it is possible to fold these reviews into other audits, they are typically performed by personnel with specialized knowledge of the health and safety regulations. The auditor should ask the auditee about the processes used to conduct health and safety compliance reviews and reviews of compliance to other, non-OHSAS 18001 requirements that the organization has adopted. If a documented procedure is available, have the auditee walk you through it. Pay attention to frequency of the audits, who performs them, how they are reported, and how any actions coming out of the review are tracked.

One of the reasons why it is recommended the evaluation of compliance to legal and other requirements be performed now, instead of folding it into the monitoring and corrective action audit, is that it fits well with the audit strategy. At this point the auditor has seen the listing of the requirements that apply and has asked for and been presented with a copy of a number of the requirements themselves. It now a simple matter to ask the auditee to provide evidence that the organization has evaluated compliance to the requirements that the auditor has in front of them. The auditee might provide me with a compliance report, completed checklists, or a consultant's evaluation. The auditor should now evaluate the thoroughness of the review. Because the auditor has the requirements, he or she can easily scan the records of the review to see if there is evidence that the requirements were evaluated. The evidence may be in the form of statements or conclusions in the compliance report or questions on the checklist, but what the auditor wants to know is whether the reviewer did a thorough job of evaluating compliance to all of the requirements. This is where management system audits differ from regulatory compliance reviews. Whereas management system audits sample compliance to important requirements, compliance reviews should verify conformance to all requirements.

If the records of the compliance review are general, stating for example, "The review verified compliance to the requirements of 19 CFR 1900–1910," then the auditor should consider generating a finding, typically in the form of an OFI that notes that the detail in the compliance records could be expanded to provide more assurance that the organization is meeting its regulatory requirements. If other components of the audit find regulatory violations then the auditor could cite the issue as an NC, using a lack of sufficient evidence of a thorough review and the auditors own findings as the basis for the NC.

Some organizations rely on outside consultants or corporate staff to conduct their compliance reviews. This is not a concern, as long as the records of the review provide sufficient evidence of a thorough compliance review.

Most internal management system auditors are not health and safety specialists and therefore may feel that they cannot evaluate the regulatory compliance sections of the specification. In fact, the auditor does not have to be a health and safety expert to do a credible job. If you note the audit strategy presented, there is nothing that requires an extensive knowledge of health and safety regulations on the auditor's part. The auditor must only ask the right questions and use the answers or evidence provided to guide him or her. The auditee will provide the list of regulations, which has already been evaluated (assuming the organization is certified), by the outside **registrar** auditor, who is knowledgeable in health and safety regulations. From that list the auditor picks several regulations and said, "Show me." The auditee then pulls down the regulations or prints off a copy for the auditor. Next the auditor asks the auditee to explain how compliance reviews are conducted and then asks for evidence that the regulations provided were a part of the compliance review. When the auditee gives the auditor copies of the reviews, he or she can easily scan the records to see if there are questions or statements relating to the regulations that have been provided. If the auditor can not find them, he or she says, "Show me." Health and safety expertise is required to do the compliance review but not to verify that the review was done.

- Were weaknesses identified in the compliance reviews acted on? If violations or weaknesses were noted in the compliance reviews, then the auditor should verify that actions were taken or have been implemented to address the deficiencies. Check the corporate policy on how to cite regulatory findings, because there may be specific actions that need to be taken when regulatory violations are found.

Emergency Preparedness and Response

The last topic of the health and safety planning audit is verification that the organization has the appropriate procedures and plans to respond to or to prevent health and safety accidents and emergencies. This is appropriate during the audit of the planning process, because this should be a planning activity and not a routine occurrence and in the United States, many of the plans and procedures are required by law.

The auditor can evaluate this area by first asking for copies of the emergency response plans the organization maintains. Common plans include the following:

- control of blood-borne pathogens;
- personal injury response plan;
- spill plans; and
- fire and evacuation plan.

Which plans are maintained by the organization is dependent on the nature of the organization's products, services, and activities. When the plans are presented, the auditor can ask the auditee to show him or her the requirements that describe what must be in the plan, where it is required by law. With the regulatory requirements in hand, the auditor can spot-check the plans for their content and currentness. Deficiencies are commonly found in these plans when compared to their regulatory requirements. Note that many of these plans focus not just on response but also on prevention. As such, the number of actual incidents or emergencies is an indirect measure of the quality of these plans.

The auditor should also verify that the plans are up to date. It is not at all uncommon to find that personnel named as emergency responders or coordinators no longer work at the facility. It does not do any good to call someone in the middle of the night during an emergency if they no longer are employed at the facility.

The auditor should ask for a list of incidents, accidents, injuries, or other emergencies that occurred since the last audit. It helps if the auditor already knows about some of these events, which is likely because this is an internal audit. The auditor should verify that reviews were held after these occurrences and that the adequacy of the response and the organization's response plans and procedures were evaluated as part of these reviews. If no evidence of any review can be provided, generate an NC.

Finally, the auditor should verify that some form of drill, test, or walkthrough was conducted to evaluate the adequacy of the procedures and plans and to familiarize those responsible for action. The organization is required to establish periodicity requirements for these tests, and the auditor should confirm they are being conducted as required. Failure to conduct the drills or tests should result in an NC.

Operational Control

Operational control implements the actions and safeguards needed to minimize any adverse health and safety risks associated with the hazards identified during health and safety planning. The major types of operational controls are engineered safeguards (guarding, interlocks, railings, etc.), administrative controls (procedures, policies, and practices), and PPE (gloves, aprons, or face shields). Good safety practice is to apply the controls in the order just presented (i.e., first apply engineered safeguards, then consider administrative controls, and rely on PPE as a complementary or last resort). The process organizations use to implement the operational controls normally looks something like that shown figure 8.4.

Figure 8.4 Operational Control Activities

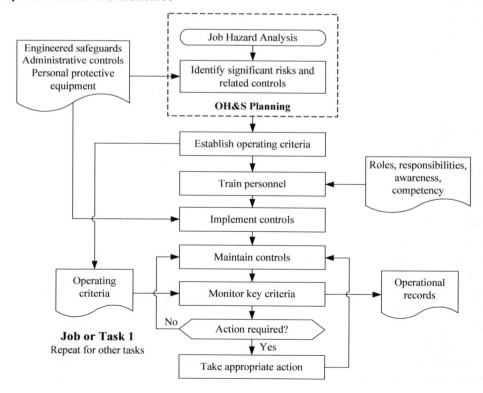

The controls themselves are normally defined as part of the health and safety planning process as described in chapter 4. Most organizations find that many of the controls are already in place, although they may not have been formalized yet. Formalizing the controls in documented procedures, JHAs, and instructions is part of the process of developing administrative operational controls. Again, remember that the preferred approach is to first apply engineered safeguards, then administrative controls, and lastly PPE.

The reader is reminded that this process would be repeated for each job or task associated with a significant health and safety risk (unacceptable, high, and moderate risk). Rather than conduct one audit of operational control, it is normally more efficient and effective to conduct multiple audits, with each series of audits focused on different processes, activities, or areas. This provides a more continuous evaluation of operational controls throughout the organization and allows easy integration with QMS and EMS audits, if desired.

Note that the evaluation of operational control will also assess the outputs from some of the key support activities, including training and awareness of employees' roles, responsibilities, and authorities. The auditor should also examine the control of any health and safety documents and records associated with the process, as well as the calibration of any health and safety monitoring devices found to be in use. As noted in the introduction to this chapter, the auditor should also verify that any monitoring of key characteristics and operating criteria is being performed as needed to maintain the health and safety controls. When performed in this fashion, the operational control audit becomes a process-based audit similar to the process-based audits conducted of an organization's QMS.

AUDIT STRATEGY AND KEY AUDIT POINTS

The auditor should begin the audit by examining the JHA for the task that he or she has been assigned to evaluate and the task's associated controls. If the auditor has been assigned to evaluate an area, then he or she should identify all of the JHAs associated with that area. Keep in mind that some tasks or activities may not have been reviewed by JHA yet. The job listing discussed in chapter 4 should provide information on which tasks have been evaluated and which have not.

Because many operational controls take the form of procedures, the auditor should obtain any procedures that relate to the control of the activities or significant hazards associated with the tasks that he or she has been assigned to audit. Examples include waste handling procedures, chemical mixing instructions, and procedures and instructions for operations of forklifts. The auditor should prepare for the audit by reviewing these procedures and modifying or developing checklists of questions and key audit points. The auditor should also make copies of the JHA for the tasks to be reviewed. If they are posted at the work site the auditor can instead review the postings on the floor.

With the JHAs that define the hazards and controls for each task to be reviewed and an understanding of how these tasks should be controlled, the auditor is now ready to conduct the review. The auditor will be talking to the operator involved in performing the task. The auditor should introduce himself or herself and set the auditee at ease and begin by asking him or her *"Are there any health or safety considerations for this task?"* The auditor is verifying operator awareness of the significant health and safety hazards of his or her job. The specification requires that operators have such an awareness, along with the consequences (i.e., injuries or damage) if they do not follow procedures.

Once the auditee replies then the auditor can ask about the effects if they are not done properly. Typical answers might be *"It could pull me into the machine"* or *"It could cause an explosion."* The auditor is ensuring that the auditee understands why it is important to control the activity. The auditor can then ask how these activities are controlled. During the auditee's response the auditor can verify that the activity is being properly controlled in accordance with company policies, procedures, and the JHA. The auditor should verify that control are in place and operating (e.g., equipment guards installed, earplugs, eye protection, or other PPE worn, and procedures and policies being followed) and should ask to see any records required to be maintained by the operator.

The auditor should verify that monitoring of key attributes of the activity is being performed as required by local procedures and practices. For example, assume a chemical mixing process specifies a maximum temperature to safely control the chemical reaction and that the operating instruction requires monitoring of the temperature every fifteen minutes. The auditor should verify that temperature monitoring is being performed through examination of the mixing records. The chemical reaction temperature is an operating criterion, and the fifteen-minute checks are monitoring requirements. If instead the reaction was instrumented with an alarm circuit if the maximum temperature was exceeded (an engineered safeguard), then the auditor would verify that the alarm circuit was operational.

The auditor should ask the operator *"Are you required to do anything if the maximum temperature is exceeded? What do you do and at what temperature?"* Now the auditor is also verifying the operator's understanding of his or her roles and responsibilities relating to control of the risk, along with knowledge and understanding of what they must do (i.e., training and competency). For specific tasks such as lockout-tagout, confined space entry, and forklift operation, the auditor may also want to jot down the auditee's name for later verification of training.

For most activities it is desirable to also observe the activity being performed. This may require careful coordination with area management, normally through the audit program manager, because these activities may not be performed all that often. Their nonroutine nature makes it all that much more important to actually observe them, if possible. Examples include lockout-tagout and permit-required confined space entry. People are normally fairly competent in the performance of routine activities; it is usually the nonroutine activities that cause problems. If the operators have specific responsibilities for emergency response relating to the activity, such as confined space rescue, the auditor can also question them on these responsibilities.

During the audit the auditor should note the revision date of any OH&S system procedures, JHAs, or instructions found in the area for later comparison to the master list of documents. He or she should also review any monitoring records for legibility and completeness. Any health and safety monitoring equipment in use in the process should be checked for calibration. He or she should also verify that proper maintenance is being performed on equipment and facilities used in the process when maintenance has been identified as a means of properly controlling the risk. This includes maintenance on any PPE, such as respirators.

If the organization has established any improvement objectives or targets relating to this activity or its hazards, then the auditor may want to ask questions to determine if the operator is aware of the objectives. Communication of improvement objectives should flow down to those most directly involved in meeting them and that is the employee most directly involved in the activity. The auditor would normally issue an OFI if he or she finds that employees are not aware of improvement objectives and targets associated with their activities or areas.

Performance Indicators

The auditor should expect to find performance indicators for the operational controls being evaluated. The auditor should review these indicators before the audit to help evaluate overall performance and effectiveness and to help focus the audit. During the audit, the auditor should discuss these indicators with area management and operators to verify awareness of health and safety performance and actions being taken to improve it. Areas of poor performance, as indicated by the indicators, should be evaluated for actions to correct the performance. The lack of any action to address significant performance issues can serve as the basis for an NC citing ineffective implementation. The following are some examples of performance indicators used to measure how well an organization is controlling its significant health and safety risks:

- Accident or injury rates in the area under evaluation,
- Near misses in the area under evaluation,

- OSHA 300 log,
- Lost workdays in the area under evaluation, and
- Accident or incident reports.

Monitoring and Corrective Action

Monitoring and corrective action focuses on measuring the performance of the HSMS and on the implementation of actions when it is not performing as it should. The reader should note that some components of monitoring have been incorporated into other audits—monitoring of regulatory compliance was reviewed during the health and safety planning audit and monitoring of operational control criteria is evaluated during the operational control audits. Likewise, corrective action for violations to regulatory requirements arising out of the compliance reviews was evaluated during the health and safety planning audit. The focus of this evaluation will be on the overall monitoring of health and safety performance and on the corrective and preventive action program.

The general process of monitoring and corrective action looks something like that shown figure 8.5. The reader will note that interrelationships with the health and safety planning, operational control, and improvement processes have been shown in the dashed boxes. While there is a possibility of some overlap between the components and therefore some redundancy between the audits, the audit does not take a large amount of time and the redundancy is minimal.

Some audit program managers may want to combine this evaluation with the evaluation of the improvement process. The improvement process includes a review of the setting of health and safety objectives, man-

Figure 8.5 Occupational Health and Safety System Monitoring and Corrective Action

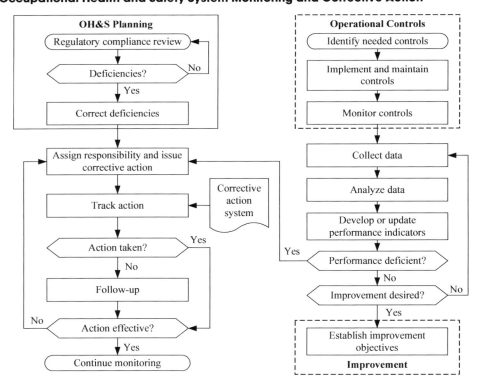

agement programs, and the conduct of the management review. There is a natural linkage between that audit and the evaluation of monitoring and corrective action. They were split into separate reviews, because the interviewees will be different—the health and safety management representative or appointee for the audit of monitoring and corrective action and executive management in the case of improvement. Even so, it may still make sense to combine the audits realizing that an audit plan may be required to support the multiple interviews.

MONITORING

The auditor should begin by asking the auditee (typically the management appointee) how overall performance of the OH&S system is monitored. The auditor should ask to see the performance measures. During his or her review, the auditor should look for trends and performance to defined goals where they have been established. The auditor should question any areas where performance is not meeting the goal or the trend is going in the wrong direction. For sustained negative trends or performance far below the goal, ask what actions are being taken to correct the performance. Failure to take action for significantly poor performance is the basis for an NC. If the period-to-period performance cannot be assessed because of changes in operations (e.g., large increases in production activity or the number of employees), then an OFI should be issued stating that monitoring could be improved through normalization of the data. Normalization means adjusting the indicator to account for the volume fluctuations, normally resulting from production activity. Keep in mind the goal of this monitoring is to evaluate health and safety performance improvement. If the indicators fail to provide useful data in this regard, they should be modified.

Because the conduct of JHA is an ongoing activity and may take an extended period to complete, the auditor should also evaluate performance in completing the analysis identified as being required during OH&S planning. Steady progress should be evident in the completion of JHA and in taking action to reduce risk. If this performance was already evaluated during the audit of OH&S planning, it need not be repeated here. On the other hand, considering the importance of this task, it would not hurt to look at it again.

Areas where performance is satisfactory but where additional improvement is desired would normally become improvement objectives. If the audit program manager has combined this audit with the improvement audit, then the auditor could flow into an evaluation of the improvement objectives now. Otherwise, the auditor would move into an evaluation of the internal audit program.

OCCUPATIONAL HEALTH AND SAFETY SYSTEM AUDITS

The auditor should ask to see the audit procedure, audit schedule, and copies of the internal audit reports performed since the last audit of the program. Verify that audits are being scheduled, conducted, and reported per the local procedure. Note that areas of significantly poor performance, evaluated just before the start of this topic, should have received additional audits or should show up on the audit schedule as additional audits to account for status of the area.

The auditor can then examine audit performance. Are audits being completed per the schedule? Probe into the reasons for significant slippages to the schedule. Insufficient numbers of trained auditors or difficulty getting part-time auditors released to perform audits may explain the inability to get the audits done. Note that some rearrangement of the audit schedule should be expected if the audit program manager is responsive to the status and importance when scheduling audits.

The auditor should compare the systematic weaknesses noted in these audit reports to external audits of the same areas or processes. If external audits found systematic weaknesses in areas where internal audits did not, then this may indicate a lack of effectiveness of the internal audit process. The internal audit must have been performed before and around the same period of time for this comparison to be valid.

The auditor can also evaluate the completeness of the audit reports and timeliness of their issuance against local requirements. The auditor may also want to verify auditor training and independence, if not known.

Audit findings should be reviewed. Auditors should evaluate whether the audit findings contain sufficient detail to fully understand the nonconformity, opportunity, or best practice. Any NC reports or CARs should get special attention. If the auditor cannot understand the issue from reading the findings, then there is a high likelihood the auditee may not either. If the auditor finds numerous poorly worded or incomplete write-ups then a finding should be generated.

The auditor should now ask to see the tracking system used to monitor the status of actions associated with audit NCs. The audit program manager may have a separate tracking system or may use the corrective action system. The auditor should check on the status of any audit NCs, evaluating whether action to correct the finding is proceeding. Any instances of overdue, outstanding corrective action should be investigated. Unless there is additional evidence of actions being taken to resolve the problem, issue an NC. Audits provide no value if action is not taken to resolve the deficiencies they find. Also confirm that verification of action taken is being performed and that audits are being used to verify the long-term effectiveness of actions taken to address nonconformities identified during previous audits.

Finally, the auditor should also confirm that the final audit file contains the information required by the local procedure. Typically this includes the audit report and any checklists used. Maintaining the checklists is not mandatory unless required by local procedures, but it is strongly recommended.

NONCONFORMANCE AND CORRECTIVE AND PREVENTIVE ACTION

Armed with information on overall OH&S system performance and the results of the internal OH&S system audits, the auditor can now evaluate the corrective and preventive action system. The auditor should ask for a copy of the corrective and preventive action procedure(s) and should verify that the system is being operated in accordance with the procedure.

The auditor should ask the program coordinator to be walked through the process, stopping to ask questions and to view evidence for key points. During the review, the auditor should verify through an evaluation of the log, database, and corrective and preventive actions that information and timing requirements mandated by local procedures are being met.

- Corrective action use: It is often found that the only time a corrective action is issued is after an audit nonconformity is received. This indicates a lack of understanding of the corrective action process or a lack of commitment to invest the time required to identify true root cause and actions to prevent recurrence. The corrective action program should be used anytime a problem or nonconformity occurs that requires a root-cause determination and action to prevent recurrence. While some isolated problems may not require corrective action, repetitive or significant problems do. Problems involving inadequate machine guarding, potential violations to regulatory requirements, and repeated disregard of the organization's PPE policies are a few examples where corrective action is warranted. Many repeat items noted during periodic safety walkthroughs also fall into this category. Taking care of these problems informally forgoes the benefits that a disciplined problem-solving methodology enacted through the corrective action process provides. Problems are likely to recur, but failure to identify them using the corrective action process makes them hard to monitor. The auditor should issue an OFI if he or she finds that all or a majority of the CARs were generated through the audit process.
- Corrective and preventive action tracking: Verify that the program coordinator is actively monitoring

the status of outstanding corrective and preventive actions. Outstanding, overdue corrective actions should be followed-up on and bumped up to more senior management (i.e., the management appointee), if follow-up is not successful in getting action to occur.

- Quality of the action taken: Did the action taken to address the problem address immediate and containment actions, any short-term or temporary actions, long-term action to eliminate the cause and prevent recurrence, and any remedial action that may have been necessary? These same action categories apply to nonaudit corrective and preventive actions.

Performance Indicators

This audit focused on the evaluation of performance indicators, examples of which were provided at the end of the section that discussed how to evaluate operational control. In addition to these indicators, the auditor may find measurements relating to the performance of the corrective and preventive action and internal audit programs. A summary of some of these indicators is provided here:

- percentage of internal audits performed on schedule;
- number of qualified internal auditors;
- percentage corrective actions recurred;
- percentage of JHAs completed, period to period;
- percentage of tasks classified as negligible or tolerable (this metric focuses on improvement in lowering overall risk through risk reduction activities); and
- number or percentage of corrective actions overdue.

Improvement

The improvement process includes the setting of health and safety objectives and targets, the establishment of management programs to ensure achievement of these objectives, and the performance of periodic management reviews. The general flow of the improvement process is shown in figure 8.6. Interrelationships with other core activities are shown in shaded boxes.

HEALTH AND SAFETY OBJECTIVES

Many companies review and set health and safety objectives during the management review process. In this case, this topic should be discussed during the evaluation of the management review. In general the audit of this topic will focus on whether objectives were set, have management programs been established for the objectives, and is the organization achieving the milestones or targets noted on the management program.

AUDIT STRATEGY AND KEY AUDIT POINTS

The auditor should ask to see the objectives and associated programs established for the current period. During this review, the auditor asks how these objectives were determined. If the auditor has recently completed the audit of the monitoring and correct action process, then he or she should be familiar with the overall performance of the HSMS system and where improvement objectives may be warranted. In particular, the auditor should verify that the views of interested parties were considered during the setting of objectives (the OHSAS

Figure 8.6   Improvement Process

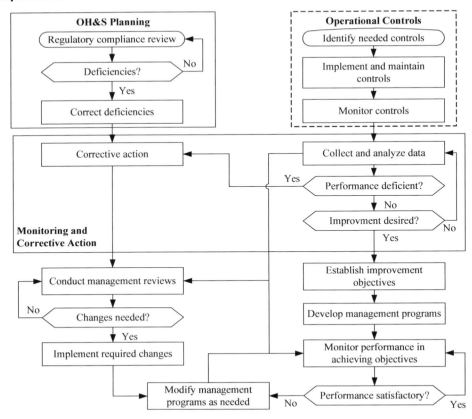

18001 specification requires this). Interested parties include outside organizations like community residents, government agencies, and the organization's stockholders and employees. The auditor should ask to see the communication log to see if there were any concerns or opinions expressed by interested parties, and if there were, that these were considered.

The auditor should compare the objectives to the management programs set up to help ensure their achievement. These management programs, or action plans, must contain details on what has to be done, who is responsible for doing them, and when they should be done. The auditor can use the actions and time frames to evaluate performance in achieving the objectives. Some form of tracking progress should be in place. If no tracking of performance in meeting objectives is evident, other than during an annual or semiannual management review, consider issuance of an OFI. Such infrequent monitoring does not allow sufficient time to recovery actions if progress is deficient.

Communication of improvement objectives is also important. The auditor may ask the auditee about the methods used to communicate objectives throughout the organization. Note that the operational control audits evaluate the effectiveness of these methods by asking employees about any objectives established for their departments or activities. The auditor may want to issue an OFI, if no formal means of communicating objectives have been established.

Of course the most important item to verify is the organization's performance in meeting its objectives. With the performance objectives and management programs in hand, the auditor should be able to establish whether the management programs have been effective in supporting achievement of the health and safety objectives. The auditor should question any areas where performance is not meeting the defined goals. For sustained negative trends or performance far below the goal, ask what actions are being taken to correct the performance. Failure to take action for significantly poor performance is the basis for an NC.

MANAGEMENT REVIEW

The purpose of the management review is to drive action. The auditor must always keep this in mind, because many organizations have evolved their management review to little more than a short status review. The management review is the key forum for management to get active and engaged in improving the management system's performance and for driving action that leads to performance improvement. Although management should demonstrate their overall commitment and support on a routine, day-in and day-out basis, the management review is somewhat like a strategic planning session, in which high-level decisions and actions on what the organization needs to do are generated and then deployed. Management reviews should be major events in the lifecycle of the 18001 management system. Unfortunately the opposite is often true—the management review is a short, one- or two-hour report out once a year that must be endured versus what should be anticipated.

The auditor will need to interview senior management during the evaluation of the management review process. If a procedure is available the auditor should use it to follow along as the manager describes the process used to conduct the review. He or she should also ask for the records from the last two management reviews conducted. The following requirements should be verified by the auditor as he or she evaluates the management review process.

- Who attends? The management review is a top-level review. Most auditors consider top management to consist of the senior person at a facility and his or her direct staff. While not everyone has to be there, a majority should attend. If the organization has a procedure that provides guidance on who should attend, then the auditor should audit to that guidance.
- How often are management reviews held? The management review must be held at planned intervals. How long should that interval be? That really depends on the maturity and status of the occupational HSMS. Most auditors consider annual reviews to be the minimum frequency. If the management system is relatively new or if performance indicators show major problems with the organization's processes or performance in achieving its objectives, then more frequent reviews are probably appropriate. If audits and performance trends indicate that the management system is in shambles and if the frequency of the management review has been left unchanged at annually, then a finding is warranted.
- How long does the review take? While there is no required duration for the management review, the auditor should use what he has learned from the evaluations of the other improvement processes to evaluate whether sufficient time is being provided to conduct a thorough review. If a global review and analysis of data was performed before the review then performance information could be summarized, shortening the time for the review. If the analysis of data will be performed as part of the review, then more time will be required. It would be hard to believe that an annual management review could be conducted in under two or three hours under any circumstances. An auditor might accept an hour-long review if it were held monthly, as is the case in some companies in which a portion of the system is evaluated at each review. Because there is no stated duration for the review and because it is hard to tie duration directly to effectiveness, auditors do not issue findings relating to duration. They should, however use the auditee's answer to evaluate how deeply to research into other aspects of the review.
- What sources of information are included in the review? The auditor should look for evidence that the key inputs into the review are evaluated. These include:
  - the results of internal and external management system audits;
  - the results of compliance reviews;

- o key performance indicators;
- o status of corrective and preventive actions;
- o follow-up actions from previous reviews;
- o changes that could affect the 18001 management system (e.g., new technologies, new processes, new or changed safety regulations, new facilities); and
- o recommendations for improvement, including changes to the OH&S policy, the organization's procedures and objectives.

Ideally the organization will have an agenda or instruction that states the items that should be included in the review. There should be solid evidence that these items were evaluated, preferably as discussion points in the records or meeting minutes from the review. If the only evidence is that they showed up on the agenda, then look for evidence of their review in the actions arising from the review or ask the auditee to summarize the results presented for these items. If all show up on the agenda, but some have no discussion in the records, then the auditor generally accepts the evidence on the agenda. If there is no evidence for these items or if the only evidence is the agenda (i.e., there are no actions or discussion of the items in the records or meeting minutes, the auditee can not remember specifics of what was discussed) then issue an NC. The standard requires that records of the management review, and any actions arising from it, be maintained.

- What actions resulted from the review? Clearly the intent of such a high-level review is to drive action. The auditor should review the records of the review to determine if actions are an output from the review. If the auditor's review of the last two management reviews fails to identify any actions or decisions arising from the review, then issue an NC or an OFI, depending on the auditor's knowledge of the system status. If the 18001 management system is now working based on audit reports, regulatory citations, and process performance data, then issue an NC. If the management system is in reasonably good shape based on these indicators, then issue an OFI. If the system is in great shape, then the auditor may choose not to issue a finding. The key to being able to generate a finding is to tie it into system status.
- How are the actions tracked? If there were actions arising from the review, then ask how these actions are tracked. Use the records from the last management review to determine if the actions arising from the review were completed or discussed during the most current review. Issue an OFI if there is no formal system for tracking actions arising from the review other than going back and looking at the review records from a year ago.
- The auditor must use ample tact and perspective when evaluating the management review process. Senior managers may support findings in other areas of the company. The auditor should not push too hard but rather point out the benefits that a robust management review should provide. Tie findings into other problems noted during previous audits. Reinforce the consequences of system failures, where known. The auditor has to make an organization want to spend the time to do a thorough review.

PERFORMANCE INDICATORS

The auditor may find some of the following metrics or indicators relating to the performance of the improvement process:

- percentage of targets achieved;
- percentage of objectives met;
- number or percentage of actions overdue;

- number of improvement projects undertaken; and
- health and safety performance gains from continual improvement activities.

Key Support Processes

The final group of activities is the key support processes. These activities support all components of the management system. They have been grouped together for convenience and to allow for an efficient audit. Portions of these activities will be evaluated during other process-based audits. These audits will evaluate only the centralized elements of these processes.

DOCUMENT CONTROL

The auditor should ask to see the document control procedure. The auditor should ask the auditee to walk through the procedure. Ask the auditee to show components of the system, such as the master list, files of archived documents, and so on as they come to them in the procedure. During the reviews of these elements the auditor should verify that they are being maintained. The auditor may want to sample select a few documents to verify that only authorized personnel reviewed or approved documents. If this is checked during other process-based audits, then this may not be necessary.

For the most part the auditor is verifying that there is a system to locate and distribute documents, there is a system that establishes the appropriate revision status and that there is a system to properly archive and safeguard obsolete documents. The auditor may also want to evaluate the backlog of document change requests to determine if resources supplied are sufficient to allow timely document preparation and issue.

The auditor should also verify proper control of health and safety regulations and assurance that there is a method to stay aware of changes to them. Many companies rely on subscription services, links to government websites where changes are announced, and corporate staff to maintain this awareness. The auditor should verify some method is in place and is being used. Note that this review may also be performed during the planning audit when evaluating legal and other requirements.

RECORD CONTROL

The auditor should ask to see the record control procedure and the records matrix. The auditor can select several records from the matrix and verify that these records are stored and retained as indicated in the matrix. Records in long-term storage should be properly indexed and protected against deterioration and loss. If the records matrix identified a retention time of a year, then the auditor could ask for a record generated six months ago to see if the records management system can retrieve the record. If record retention is decentralized this may not be practical. If records are stored electronically, the auditor should verify that records are properly backed up and are password protected. Finally, if the organization has a record disposal policy, then the auditor may want to verify that records are being disposed of as required by local procedures.

Keep in mind that many of the records that must be maintained have legal retention requirements. The auditor may want to have the auditee show them the legal retention requirement for the records if doubt exists.

TRAINING

Like many other activities discussed in this section, training and competency should be evaluated during each and every process-based audit. The focus of this evaluation is on the centralized training functions.

Especially important is the process for identifying competency requirements. Auditors may find that personnel have met all of the defined competencies and have training records to prove it, but if the critical competencies are not correctly defined in the first place, then the result will be poor performance, nonconformities, and potentially regulatory violations. It should also be noted that competency identification is not a one-time evolution; competency requirements change over time as new methods and technologies are adapted by the organization. For this reason the audit of the centralized training activities focuses on the identification and communication of required competencies. Note that competencies required for specific tasks may be identified on the work-specific JHAs and associated training records maintained at the work site.

Another area in which training programs are often deficient is in the verification of training effectiveness. Although tests, pre- and post-training quizzes, and student feedback forms provide some information on the effectiveness of training delivery, the real test is whether personnel can properly do their job. This requires actual observation and ongoing monitoring. Training departments can and should use the results of internal audits, internal and external problem reports, and post-training feedback to evaluate how effective the training program is.

The auditor can begin the evaluation by asking to see a copy of the training procedure, if one has been developed. Most procedures are laid out along the natural flow of the process, which in this case would be identification of competency requirements, delivery of training, and verification of training effectiveness. The auditor should ask to be walked though the process, paying particular attention to the items discussed here.

- Competency requirements: The auditor should ask to see the competency matrix or its equivalent that identifies the competencies that must be achieved. Some organizations use job descriptions to identify competency requirements. The auditor should use his or her knowledge of the tasks performed within the organization to evaluate several functions for review. It helps if the audit program manager selects more experienced auditors to do this evaluation, because they will have a better idea of the types of competencies required to perform functional tasks as a result of their previous audits. As an alternative, this review could be added to other audit checklists. Although it would add to the length of time required to perform these audits, it would result in a more thorough and penetrating evaluation, because the auditor would be familiar with the types of competencies required, having just evaluated the process. It would also result in better coverage than could be attained from a sampling of functional duties performed here.

- The auditor should also ask how competency requirements are updated and maintained. In the dynamic business environment of today, new technologies, new materials, and new processes are constantly being introduced. Some method of reevaluation and adjustment of competencies should be in place to respond to these dynamics. The training matrices or job descriptions should be controlled documents, and the auditor should be able to identify updates and modifications. If the training matrices have been in use for some time and have never been updated, then the auditor should probe this area. The auditor should verify that some method is in use, otherwise issue an OFI.

  Finally, note that identification of required competencies requires the support of the various department and process owners who know the tasks best. Human resources really cannot do a credible job without their input. Some type of formal or informal task analysis is required. Because so many different owners are involved in determining competency requirements, there is often a wide variance in the level of detail and quality of the output (the competency requirements). Auditors should be sensitive to this variation and should note functions where it does not look like much effort was put into defining the required competencies.

- Competency attainment: The next task is to verify that adequate progress is being made to attain the mandatory competencies. The auditor can ask how competency attainment is monitored. Competency attainment may be indicated on the competency matrix or on an individual's training plans. If the organization has criteria for training, such as *"basic required skills must be attained during the initial three-month probationary period*," then the auditor can sample several recently hired personnel (between three and six months ago) to see if the policy is being followed. Normally the auditor will select records of personnel hired since the training procedure or program was established. Employees who were working at the organization before that time are often grandfathered based on their previous demonstrated skills or experience. This is acceptable as long as there is some evidence that some review of their current skills against the newly identified competencies was performed, and they were not just signed off in mass as being competent. The evidence, if not in writing, may be in the form of an interview with the functional head of the department responsible for identifying and verifying current competencies.

The auditor is now looking to see if competency gaps have been identified and if actions to close those gaps are being taken, normally through the scheduling of training on an individual's training plan. The auditor is also indirectly monitoring the deployment of resources in support of training. In today's lean and cost-conscious business environment, it is not uncommon to find the first thing cut is the training budget. Failure to meet established training goals or recurring delays in providing required training resulting from resource limitations should be noted as findings. Delays in providing optional training can be justified; delays in providing mandatory training cannot. If the auditor finds that mandatory training (including any required by safety regulations) is not being provided, then he or she should inquire what additional safeguards are being taken to protect the employee pending provision of the training. The auditor is primarily looking for trends and broad issues with competency attainment, because the specific training and competency of employees should be reviewed during other audits.

- Verification of training effectiveness: Training verification should verify that employees can actually do their jobs as a result of the training. Most verification instead evaluates training delivery and is highly biased toward the presentation skills of the trainer, the quality of the visual aids, or the trainer's skill in prepping the trainees for the post-training quiz. Effective verification normally requires observation of the individual performing the task, evaluation of indicators that they can or cannot perform the task correctly, or feedback from the individual *after* they have tried to apply the training on the job. Direct observation works well for many tasks.

During the review of training the auditor must be aware of the additional consideration that some organizations have a legal obligation to provide certain types of training, such as hazardous waste training for employees who handle hazardous wastes, training in the proper use of PPE for those who must use respirators, and forklift driver training. The auditor may want to question the training coordinator how these mandatory training requirements were identified, in addition to verifying the training was provided.

The auditor should also ensure that contract and agency personnel are made aware of the organizations OH&S policies and practices, to the extent they apply to the work they are involved with.

CONTROL OF PURCHASED PRODUCTS AND CONTRACTORS (PURCHASING)

In the 18001 management system, organizations should be most concerned with two things relating to their suppliers and contractors. First, they need to ensure that any relevant health and safety policies and requirements are communicated to suppliers and contractors. This can be done through a pass down of the requirements (e.g., pass down of any safety rules and requirements) or by briefings before the contractor commences work on-site. Most organizations have a fairly robust process to do this. The auditor should verify that the process is being followed during the audit.

The second consideration, more often lacking, is obtaining information from the supplier or contractor regarding any significant health and safety hazards arising from their activities and products. For suppliers of materials and products, this is often accomplished through receipt of the MSDS from the supplier. The auditor should ensure that the MSDSs for new products are reviewed for any health and safety concerns that may need to be addressed before using the new chemical or material. Note that there are legal requirements for this hazard communication. For contractors, a health and safety survey is often used. The survey would ask questions regarding the types of materials and chemicals to be used, how these materials and chemicals will be controlled, and the types of safety zones or controls that will be required during the activity. The auditor should verify that methods are in place to address these two considerations and are being used.

ROLES AND RESPONSIBILITIES

Roles and responsibilities connected with defined activities associated with health and safety hazards will be evaluated during operational control audits. In this audit, the auditor should focus on the definition, communication, and awareness of roles, responsibilities, and authorities of key individuals and functions needed to support the HSMS. These include the management representative or appointee, internal HSMS auditors, emergency response coordinators, safety committee members, and the CFT assigned responsibilities for JHA and determining their risks. If these roles are examined during other audits they do not need to be reviewed here.

A unique element of the OHSAS 18001 specification is the requirement to designate a management appointee. This position is clearly designed to place accountability and authority for the HSMS squarely in the lap of senior management and to engage their involvement along with their support. The auditor should verify that a management appointee has been designated, that it is being filled by a senior executive, and that this executive is involved in the operation, maintenance, and improvement of the HSMS.

COMMUNICATIONS

The final evaluation examines the process used to communicate important health and safety information within the organization and to outside interested parties. Communication of key information internally, such as roles and responsibilities, awareness, and improvement objectives, will be continually evaluated during other audits. A unique element of OHSAS 18001, not found in ISO 14001 or ISO 9001, is the requirement for employee consultation. The specification requires that employees be involved in the development and review of policies and procedures used to manage risks, that they be consulted when there are changes that affect workplace health and safety, that they be represented on health and safety matters, and that they be informed of who their HSMS representatives are and who the management appointee is.

For an OHSAS 18001 system, therefore, it is critical that there be a process to communicate concerns, issues, and improvement suggestions up the ladder. The auditor may find that no formal system exists for upward communication from employees. This will damage the support for the HSMS and will result in losing ideas that could significantly improve health and safety performance. Many organizations use their safety committees to meet the requirements for consultation. This works well, as long as the safety committee members adequately represent the workforce, are consulted as the specification requires, and understand their responsibilities to act as employee representatives. The auditor should issue a finding if no formal system exists to support upward communication or consultation from the workforce. The auditor can test the effectiveness of the system, if it exists, by asking to see the log or file of concerns, issues, or suggestions provided by employees. A blank file or log tells an auditor something is not working.

Another focus of this evaluation is on external communication. OHSAS 18001 requires that a system for communication with interested parties exits. Interested parties for a health and safety system might include state and local government officials, the local fire department, hospitals and other emergency planners, and insurance providers. This is almost universally met by the maintenance of some type of communication log.

The auditor should ask about the methods used to respond to outside parties, including who receives them, how are they handled, and who responds. The auditor should also ask to see the log. Although it may be that there are no entries in the log, indicating no outside inquiries, the auditor should verify that a system is in place to handle them if any come in.

PERFORMANCE INDICATORS

Performance indicators for document control, record control, and training are provided in this chapter. Performance indicators for the communication process may include:

- number of health and safety improvement suggestions received from employees;
- number of improvement suggestions per employee;
- number of improvement suggestions implemented;
- health and safety performance gains from employee improvement suggestions; and
- average response cycle time for outside inquiries.

Summary

In this chapter I described the strategy for an audit of an occupational HSMS. The focus was on strategies and questions for ensuring not just compliance but also effectiveness. This chapter can be used with chapter 5 and the checklists on the CD to create a robust audit process.

The next chapter will examine voluntary programs and tools that can help the organization improve its health and safety performance.

CHAPTER 9

OSHA's Voluntary Protection Program

The VPP was created by OSHA in 1982 in partnership with industry and labor. The goal of the program was to create a framework for an HSMS that would result in significant improvements in worker health and safety through the cooperative efforts of managers, employees, and their representatives by going beyond basic compliance with basic OSHA standards. The program is entirely voluntary and represents a partnership between the organization and OSHA.

The VPP process emphasizes holding managers accountable for worker safety and health, the continual identification and elimination of hazards, and the active involvement of employees in their own protection. These criteria work for the full range of industries, union and non-union, and for employers, large and small, private and public. The VPP places significant reliance on the cooperation and trust inherent in partnership.

The VPP program is progressive, offering progressive levels of achievement. Sites qualifying for VPP attain *Star*, *Merit*, or *Demonstration* status. Star participants meet all VPP requirements. Merit participants have demonstrated the potential and willingness to achieve Star status, but some aspects of their programs need improvement. Demonstration participants test alternative ways to achieve safety and health excellence that may lead to changes in VPP criteria.

Statistical evidence for VPP's success is impressive. Consistently over its twenty-year history, the average VPP work site has had an incidence rate for days away from work, restricted work activity, or job transfer that is at least 50 percent below the average for its industry.

In VPP,

- Management commits to operating an effective occupational HSMS characterized by four basic elements: management leadership and employee involvement, work site analysis, hazard prevention and control, and safety and health training. Employees agree to participate in the program and work with management to ensure a safe and healthful workplace.
- The site submits an application to OSHA that describes its system of worker protection.
- OSHA evaluates the application. If OSHA accepts it, the agency then conducts an on-site review to verify that the HSMS meets VPP requirements. With approval comes OSHA's public recognition of the applicant's exemplary HSMS.
- OSHA also periodically reevaluates the participant to confirm its continuing qualification for VPP. On-site evaluations are every two-and-a-half to five years for Star, twelve to eighteen months for Demonstration, and eighteen to twenty-four months for Merit.
- OSHA removes VPP participants from its programmed inspection lists.
- OSHA enforcement personnel will investigate workplace complaints, any fatality or catastrophe, and other significant events. After such events, VPP personnel may also review a participant's continuing eligibility for VPP.

OHSAS 18001 can be used as a model for a VPP HSMS. As will be shown, the components of 18001 address most of the requirements of the VPP or provide a foundation for its requirements. Many organizations therefore use OHSAS 18001 as the natural starting point for their participation in VPP, because 18001's structure and format is easier to understand and implement. Once the 18001 system is in place, then the VPP components can be added into the existing elements. In addition, the VPP requires the organization to have, or be able to achieve, a lost workday rate less than the industry average. In this case an organization must have a plan for how it will achieve reductions below its industry average. The implementation of an 18001 management system could be this plan.

The remainder of this chapter presents the major categories of OSHA's VPP, as presented in its self-assessment checklist and application requirements, along with 18001's corresponding clauses that could be used to address the VPP requirements. The purpose is to illustrate the alignment between the two programs and how 18001 can be used to drive VPP achievement. Readers may want to review previous chapters of this book that discuss the design aspects of the 18001 HSMS to reinforce the synergy between VPP and 18001. Comments will be included in instances where the 18001 requirements do not address a VPP requirement or where additional controls, beyond the minimum prescribed by 18001, would need to be implemented.

Note that the VPP self-assessment checklist is included in the OSHA VPP file included on the CD.

Summary

This chapter addresses the use of OHSAS 18001 to support the application for, and compliance to, OSHA's VPP, a program that focuses on going beyond basic compliance to standards and regulations. Together these systems provide a framework for a robust, effective occupational health and safety program that can provide for a safe, healthy, and effective program that will contain or lower the organization's overall health care costs and improve employees overall motivation and satisfaction.

While not specifically addressed by either 18001 or OSHA's VPP, other even more proactive components can and should be initiated as the organization matures its health and safety program. Examples include employee wellness programs, fitness programs, nutritional programs, and initiatives that include workers' families and home life. The implementation of a total health and safety program, embedded within an OHSAS 18001 management system, will go far toward making the organization a place where employees desire to work, which is a strategic imperative in today's era of the knowledge worker. Employees are truly the organization's most valuable asset; keeping them safe, healthy, and productive should be the goal of every organization.

Figure 9.1   Voluntary Protection Program to Occupational Health and Safety Assessment Series 18001 Matrix

VPP Requirement or Goal	OHSAS 18001 Requirements and Comments
Management Leadership and Employee Involvement	
A managerial commitment to worker safety and health protection.	Clause 4.2, OH&S Policy, developed and approved by top management.
Top site management's personal involvement.	Clause 4.4.1, appointment of a top manager/executive as management appointee and 4.6, management review. Note top management would also be expected to be involved in the development of objectives and management programs (clauses 4.3.3 and 4.3.4).
A system in place to address safety and health issues/concerns during overall management planning/purchasing/contracting.	Clause 4.2, OH&S Policy, all of clause 4.3, Planning, and clause 4.4.1.c for purchasing and contracting.
Safety and health management integrated with your general day-to-day management system.	Not directly addressed by 18001, but implied throughout standard and this book through the management program elements and system of reviews.

A written safety and health management system—often referred to as a safety and health manual with policy and procedures specific to your site—appropriate for your site's size and your industry that addresses all the elements in this checklist.

Clause 4.4.4, Documentation. Note that 18001 does not require written procedures for each system element, while VPP does for all components of the VPP. System designers may have to develop some additional documentation to meet VPP requirements.

A safety and health policy communicated to and understood by employees.

Clause 4.2, OH&S Policy.

Safety and health management system goals and results-oriented objectives for meeting those goals.

Clause 4.3.3 Objectives and 4.3.4, OH&S Management Programs.

Clearly assigned safety and health responsibilities with documentation of authority and accountability from top management to line supervisors to site employees.

Clause 4.4.1, Structure and Responsibility, noting that the specific line of authority and accountability from site employees through supervision to top management must be shown.

Necessary resources to meet responsibilities, including access to certified safety and health professionals, other licensed health care professionals, and other experts, as needed.

Clause 4.4.1, Structure and Responsibility, noting that 18001 requires that resources be provided, not necessarily specifying access to health care professionals. This specific requirement of VPP would have to be addressed in the 18001 system.

Selection and oversight of contractors to ensure effective safety and health protection for all workers at the site.

Clause 4.4.6.c, noting additional requirements on the evaluation and selection of contractors would need to be added to the 18001 system.

At least three ways employees are meaningfully involved in activities and decision making that impact their safety and health.

Clause 4.4.3, Consultation and Communication, specifically employee involvement in the development and review of policies and procedures, changes that affect workplace health and safety and clause 4.3.3, Objectives, regarding consideration of interested parties (employee) input when setting OH&S objectives.

Annual safety and health management system evaluations on VPP elements in a narrative format, recommendations for improvements, and documented follow-up.

Clause 4.6, Management Review and 4.5.4, Audit.

Formal signed statements from all collective bargaining agents indicating support of your application to VPP.

Not addressed by 18001. This would need to be added to the 18001 system but aligns well with clause 4.4.3, Consultation and Communication.

Where no collective bargaining agent is authorized, written assurance by management that employees understand and support VPP participation.

Not addressed by 18001. This would need to be added to the 18001 system but aligns well with clause 4.4.3, Consultation and Communication.

Work Site Analysis

A baseline hazard analysis identifies and documents common hazards associated with your site, such as those found in OSHA regulations, building codes, and other recognized industry standards and for which existing controls are well known.

Clause 4.3.1, Planning for Hazard Identification, Risk Assessment and Risk Control and clause 4.3.2, Legal and Other Requirements.

Documentation within the baseline hazard analysis of your sampling strategy to identify health hazards and accurately assess employees' exposure, including duration, route, frequency of exposure, and number of exposed employees.

Clause 4.3.1, Planning for Hazard Identification, Risk Assessment and Risk Control and clause 4.3.2, Legal and Other Requirements.

Hazard analysis of routine jobs, tasks, and processes that identifies uncontrolled hazards and leads to hazard elimination or control.

Clause 4.3.1, Planning for Hazard Identification, Risk Assessment and Risk Control and clause 4.3.2, Legal and Other Requirements.

Figure 9.1   (Continued)

VPP Requirement or Goal	OHSAS 18001 Requirements and Comments
Work Site Analysis	
Hazard analysis of significant changes, including non-routine tasks, new processes, materials, equipment, and facilities, that identifies uncontrolled hazards prior to the activity or use and leads to hazard elimination or control.	Clause 4.3.1, Planning for Hazard Identification, Risk Assessment and Risk Control and clause 4.3.2, Legal and Other Requirements.
Samples, tests, and analyses that follow nationally recognized procedures.	Not specifically addressed by 18001, would need to be incorporated into 18001 system.
Self-inspections, conducted by trained staff with written documentation and hazard correction tracking, that cover the entire site at least quarterly (weekly for construction).	Clause 4.5.1, Performance Measurement and Monitoring, and clause 4.5.4, Audit. Would need to ensure minimum quarterly frequency cited by VPP met in 18001 system.
A written hazard reporting system that enables employees to report their observations or concerns to management without fear of reprisal and to receive timely responses.	Clause 4.4.3, Consultation and Communication. Note that the VPP requirements are more specific than those in 18001, and would have to be addressed in system.
Accident/incident investigations conducted by trained staff. Written findings that aim to identify all contributing factors.	Clause 4.5.2, Accidents, Incidents, Non-conformances and Corrective and Preventive Action. Note VPP requirement for training of accident investigators.
A system that analyzes injury, illness, and related data—including inspection results, observations, near-miss and incident reporting, first aid, and injury and illness records—to identify common causes and needed corrections in procedures, equipment, or programs.	Clause 4.5.1, Performance Measurement and Monitoring, and clause 4.6, Management Review.
Hazard Prevention and Control	
An effective system for eliminating or controlling hazards. This system emphasizes engineering solutions that provide the most reliable and effective protection. It may also utilize, in preferred order, administrative controls that limit daily exposure, such as job rotation; work practice controls, such as rules and work practices that govern how a job is done safely and healthfully; and personal protective equipment. All affected employees must understand and follow the system.	Clause 4.3.1, Planning for Hazard Identification, Risk Assessment and Risk Control and clause 4.4.6, Operational Control. The last requirement regarding employee understanding is addressed by clause 4.4.2, Training, Awareness and Competence.
A system for tracking hazard correction. It includes documentation of how and when hazards are identified, controlled or eliminated, and communicated to employees.	Clause 4.3.1, Planning for Hazard Identification, Risk Assessment and Risk Control and clause 4.4.2, Training, Awareness and Competence.
A written preventive/predictive maintenance system that reduces safety-critical equipment failures and schedules routine maintenance and monitoring.	Clause 4.4.6, Operational Control.
An occupational health care program appropriate for your workplace. It includes, at a minimum, nearby medical services, staff trained in first aid and CPR, and hazard analysis by licensed health care professionals as needed.	Clause 4.4.1, Structure and Responsibility (resources) and clause 4.4.7, Emergency Preparedness and Response.
A consistent disciplinary system that operates for all employees—including supervisors and managers—who disregard the rules.	Not specifically addressed by 18001. Would need to be added to the 18001 system.

Written plans to cover emergency situations, including emergency and evacuation drills for all shifts.	Clause 4.4.7, Emergency Preparedness and Response.

Safety and Health Training

Training for managers and supervisors that emphasizes safety and health leadership responsibilities.	Clause 4.4.2, Training, Awareness and Competence.
Training for all employees on the site's safety and health management system, hazards, hazard controls in place, and the VPP.	Clause 4.4.2, Training, Awareness and Competence. Training on the VPP would have to be added to the 18001 training program.
Training that enables employees to recognize hazardous conditions and understand safe work procedures.	Clause 4.4.2, Training, Awareness and Competence. Hazard recognition training for all employees may need to be added to the 18001 training program.
A method for assessing employee comprehension and training effectiveness.	Clause 4.4.2, Training, Awareness and Competence and clause 4.5.4, Audit. The audit process can be used to evaluate the effectiveness of the employee training and awareness program.
Documentation of all training that individual employees receive.	Clause 4.4.2, Training, Awareness and Competence.

VPP Application Requirement	*OHSAS 18001 Requirements and Comments*

Management Leadership

1. Commitment Attach a copy of your top level safety policy specific to your facility. Note: Management must clearly demonstrate commitment to meeting and maintaining the requirements of the VPP.	Clause 4.2, OH&S Policy. Note that a commitment to comply with VPP requirements should be called out in the policy statement.
2. Organization Describe how your company's safety and health function fits into your overall management organization. Attach a copy of your organization chart.	4.4.1, Structure and Responsibility and 4.4.4, Documentation. The VPP requirement should be addressed in the organization's policy manual, which should include a copy of the organization chart referenced here.
3. Responsibility Describe how your line and staff are assigned safety and health responsibilities. Include examples of specific responsibilities.	4.4.1, Structure and Responsibility and 4.4.4, Documentation. Describe in the organization's policy manual or in separate job descriptions/documentation.
4. Accountability Describe your accountability system used to hold managers, line supervisors, and employees responsible for safety and health. Examples are job performance evaluations, warning notices, and contract language. Describe system documentation.	4.4.1, Structure and Responsibility and 4.4.4, Documentation. Add this section to the organization's policy manual.
5. Resources Identify the available safety and health resources. Describe the safety and health professional staff available, including appropriate use of certified safety professionals (CSP), certified industrial hygienists (CIH), other licensed health care professionals, and other experts as needed, based on the risks at your site. Identify any external resources (including corporate office and private consultants) used to help with your safety and health management system.	4.4.1, Structure and Responsibility and 4.4.4, Documentation. Describe in the organization's policy manual.

Figure 9.1   (Continued)

VPP Application Requirement	OHSAS 18001 Requirements and Comments
Management Leadership	

6. Goals and Planning Identify your annual plans that set specific safety and health goals and objectives. Describe how planning for safety and health fits into your overall management planning process.	Clause 4.3.3 Objectives, 4.3.4, OH&S Management Program, and 4.6, Management Review.
7. Self-Evaluation Provide a copy of the most recent annual self-evaluation of your safety and health management system. Include assessments of the effectiveness of the VPP elements listed in these application guidelines, recommendations for improvement, assignment of responsibility, and documentation of action items completed. Describe how you prepare and use the self-evaluation.	Clause 4.5.4, Audit. Include the self-assessment in the audit program and on the audit schedule.

Employee Involvement	

8. Three Ways List at least three meaningful ways employees are involved in your safety and health management system. Provide specific information about decision processes that employees impact, such as hazard assessment, inspections, safety and health training, and/or evaluation of the safety and health management system.	Clause 4.4.3, Consultation and Communication, specifically employee involvement in the development and review of policies and procedures, changes that affect workplace health and safety and clause 4.3.3, Objectives, regarding consideration of interested parties (employee) input when setting OH&S objectives.
Describe how you notify employees about site participation in the VPP, their right to register a complaint with OSHA, and their right to obtain reports of inspections and accident investigations upon request. (Various methods may include new employee orientation; Intranet or e-mail, if all employees have access; bulletin boards; tool box talks; or group meetings.)	Clause 4.4.3, Consultation and Communication, and clause 4.4.2, Training, Awareness and Competence.
10. Contract Workers' Safety Describe the process used for selecting contractors to perform jobs at your site. Describe your system for ensuring that all contract workers who do work at your site are provided the same healthful working conditions and the same quality protection as your regular employees.	Clause 4.4.1.c, noting additional requirements on the evaluation and selection of contractors would need to be added to the 18001 system.
11. Site Map Attach a site map or general layout.	Not specifically addressed by 18001.

Work Site Analysis	

1. Baseline Hazard Analysis Describe the methods used for baseline hazard analysis to identify hazards associated with your specific work environment, for example, air contaminants, noise, or lead. Identify the safety and health professionals involved in the baseline assessment and subsequent needed surveys. Explain any sampling rationale and strategies for industrial hygiene surveys if required.	Clause 4.3.1, Planning for Hazard Identification, Risk Assessment and Risk Control and clause 4.4.6, Operational Control.

2. Hazard Analysis of Routine Jobs, Tasks, and Processes
Describe the system used for examination and analysis of safety and health hazards associated with routine tasks, jobs, processes, and/or phases. Provide some sample analyses and any forms used. You should base priorities for hazard analysis on historical evidence, perceived risks, complexity, and the frequency of jobs/tasks completed at your work site. In construction, the emphasis must be on special safety and health hazards of each craft and phase of work.

Clause 4.3.1, Planning for Hazard Identification, Risk Assessment and Risk Control and clause 4.4.6, Operational Control.

3. Hazard Analysis of Significant Changes
Explain how, prior to activity or use, you analyze significant changes to identify uncontrolled hazards and the actions needed to eliminate or control these hazards. Significant changes may include nonroutine tasks and new processes, materials, equipment, and facilities.

Clause 4.3.1, Planning for Hazard Identification, Risk Assessment and Risk Control and clause 4.4.6, Operational Control.

4. Self-Inspections
Describe your work site safety and health routine general inspection procedures. Indicate who performs inspections, their training, and how you track any hazards through to elimination or control. For routine health inspections, summarize the testing and analysis procedures used and qualifications of personnel who conduct them. Include forms used for self-inspections.

Clause 4.5.1, Performance Measurement and Monitoring, and clause 4.5.4, Audit.

5. Employee Reports of Hazards
Describe how employees notify management of uncontrolled safety or health hazards. Explain procedures for follow-up and tracking corrections.
An opportunity to use a written form to notify management about safety and health hazards must be part of your reporting system.

Clause 4.4.3, Consultation and Communication, and clause 4.5.2, Accidents, Incidents, Non-conformances and Corrective and Preventive Action

6. Accident and Incident Investigations
Describe your written procedures for investigation of accidents, near-misses, first-aid cases, and other incidents. What training do investigators receive? How do you determine which accidents or incidents warrant investigation? Incidents should include first-aid and near-miss cases. Describe how results are used.

Clause 4.5.2, Accidents, Incidents, Non-conformances and Corrective and Preventive Action, clause 4.4.2, Training, Awareness and Competence and clause 4.4.7, Emergency Preparedness and Response.

7. Pattern Analysis
Describe the system you use for safety and health data analysis. Indicate how you collect and analyze data from all sources, including injuries, illnesses, near-misses, first-aid cases, work order forms, incident investigations, inspections, and self-audits.
Describe how results are used.

Clause 4.5.1, Performance Measurement and Monitoring, noting that pattern analysis would need to be specifically addressed in the procedure for this clause.

Figure 9.1　(Continued)

VPP Application Requirement	*OHSAS 18001 Requirements and Comments*
Hazard Prevention and Control	

1. Engineering Controls
Describe and provide examples of engineering controls you have implemented that either eliminated or limited hazards by reducing their severity, their likelihood of occurrence, or both. Engineering controls include, for example, reduction in pressure or amount of hazardous material, substitution of less hazardous material, reduction of noise produced, fail-safe design, leak before burst, fault tolerance/redundancy, and ergonomic design changes.

Although not as reliable as true engineering controls, this category also includes protective safety devices such as guards, barriers, interlocks, grounding and bonding systems, and pressure relief valves to keep pressure within a safe limit.

Clause 4.3.1, Planning for Hazard Identification, Risk Assessment and Risk Control and clause 4.4.6, Operational Control.

2. Administrative Controls
Describe ways you limit daily exposure to hazards by adjusting work schedules or work tasks, for example, job rotation.

Clause 4.3.1, Planning for Hazard Identification, Risk Assessment and Risk Control and clause 4.4.6, Operational Control.

3. Work Practice Controls
Describe and provide examples of your work practice controls. These include, for example, workplace rules, safe and healthful work practices, specific programs to address OSHA standards, and procedures for specific operations. Identify major technical programs and regulations that pertain to your site, such as lockout-tagout, process safety management, hazard communication, machine guarding, and fall protection.

Clause 4.3.1, Planning for Hazard Identification, Risk Assessment and Risk Control, clause 4.3.2, Legal and Other Requirements and clause 4.4.6, Operational Control.

4. Personal Protective Equipment
Describe and provide examples of required personal protective equipment your employees use.

Clause 4.3.1, Planning for Hazard Identification, Risk Assessment and Risk Control, clause 4.3.2, Legal and Other Requirements and clause 4.4.6, Operational Control.

5. Safety and Health Rules
Describe your general safety and health rules. Demonstrate that there is a disciplinary system for equitably enforcing these rules for managers, supervisors, and employees.

The policies, practices and administrative controls established as part of the 18001 system form the basis for addressing this portion of the VPP application. Note that disciplinary systems are not addressed by 18001 and would need to be added.

6. Preventive/Predictive Maintenance
Describe your written system for monitoring and maintaining workplace equipment to predict and prevent equipment breakdowns that may cause hazards. Provide a brief summary of the type of equipment covered.

Clause 4.4.6, Operational Control (maintenance).

7. *Occupational Health Care Program*
Describe your on-site and off-site medical service and physician availability. Explain how you utilize the services of licensed occupational health care professionals. Indicate the coverage provided by employees trained in first aid, CPR, and other paramedical skills, their training, and available equipment.

Not specifically addressed by 18001. Would need to be added to the 18001 system.

8. *Emergency Preparedness*
Describe your emergency planning and preparedness system. Provide information on emergency drills and training, including evacuations.

Clause 4.4.7, Emergency Preparedness and Response and clause 4.4.2, Training, Awareness and Competence.

Safety and Health Training

Describe the formal and informal safety and health training provided for managers, supervisors, and employees. Identify training protocols, schedules, and information provided to supervisors and employees on programs such as hazard communication, personal protective equipment, and handling of emergency situations. Describe how you verify the effectiveness of the training given.

Clause 4.4.2, Training, Awareness and Competence.

Assurances

1. *Compliance*
You will comply with the *Occupational Safety and Health Act* (*OSH Act*) and correct in a timely manner all hazards discovered through self-inspections, employee notification, accident investigations, OSHA on-site reviews, process hazard reviews, annual evaluations, or any other means. You provide effective interim protection, as necessary. Federal applicants also agree to comply with *Title of the Code of Federal Regulations* (*CFR*), Part 1960—Basic Program Elements for Federal Employees.

Clause 4.3.1, Planning for Hazard Identification, Risk Assessment and Risk Control, clause 4.3.2, Legal and Other Requirements, clause 4.5.1 Performance Measuring and Monitoring (monitoring compliance (legislation and regulatory requirements), clause 4.5.2, Accidents, Incidents, Non-conformances and Corrective and Preventive Action, and clause 4.5.4, Audit.

2. *Correction of Deficiencies*
Within 90 days, you will correct safety and health deficiencies related to compliance with OSHA requirements and identified during any OSHA on-site review.

Clause 4.3.2, Legal and Other Requirements, and clause 4.5.2, Accidents, Incidents, Non-conformances and Corrective and Preventive Action. This would need to be made a part of the corrective action process.

3. *Employee Support*
Your employees support the VPP application. At sites with employees organized into one or more collective bargaining units, the authorized representative for each collective bargaining unit must either sign the application or submit a signed statement indicating that the collective bargaining agent(s) support VPP participation. OSHA must receive concurrence from all such authorized agents to accept the application. At non-union sites, management's assurance of employee support will be verified by the OSHA on-site review team during employee interviews.

Clause 4.4.3, Consultation and Communication.

4. *VPP Elements*
VPP elements are in place, and management commits to meeting and maintaining the requirements of the elements and the overall VPP.

Clause 4.2, OH&S Policy, clause 4.5.4, Audits and clause 4.6, Management Review.

Figure 9.1   (Continued)

VPP Application Requirement	OHSAS 18001 Requirements and Comments
Assurances	
5. Orientation Employees, including newly hired employees and contract employees, will receive orientation on the VPP, including employee rights under VPP and under the *OSH Act* or *29 CFR* 1960.	Clause 4.4.2, Training, Awareness and Competence. The VPP would need to be added to orientation and awareness training.
6. Non-Discrimination You will protect employees given safety and health duties as part of your safety and health management system from discriminatory actions resulting from their carrying out such duties, just as Section 11(c) of the *OSH Act* and *29 CFR* 1960.46(a) protect employees who exercise their rights.	Not addressed by 18001. Would need to be added to the 18001 system.
7. Employee Access Employees will have access to the results of self inspections, accident investigations, and other safety and health data upon request. At unionized construction sites, this requirement may be met through employee representative access to these results.	Clause 4.4.3, Consultation and Communication.
8. Documentation You will maintain your safety and health management system information and make it available for OSHA review to determine initial and continued approval to the VPP. This information will include: ☐ Any agreements between management and the collective bargaining agent(s) concerning safety and health. ☐ All documentation enumerated under Section III.J.4. of the July 24, 2000 *Federal Register* Notice. ☐ Any data necessary to evaluate the achievement of individual Merit or 1-Year Conditional Star goals.	Clause 4.4.4 Documentation, clause 4.4.5, Document and Data Control, and clause 4.5.3, Records and Records Management.

Further Reading

Books and Articles

Kaplan, Robert and David Norton. *The Balanced Scorecard.* Boston: Harvard Business School Press, 1996.
Swartz, George. *Job Hazard Analysis: A Guide to Identifying Risks in the Workplace.* Rockville, MD: Government Institutes, 2001.

Standards and Specifications

ISO 14001. *2004 Environmental Management Systems: Requirements with Guidance for Use.* Milwaukee, Wis.: American Society for Quality, 2004.
British Standards Institute. *OHSAS 18001:1999 Occupational Health and Safety Management Systems: Specification.* London, UK: British Standards Institute, 2002.

Websites

The OSHA website, maintained by the U.S. Department of Labor, provides useful safety and health tools and training materials for a variety of industries. Visit this site at http://www.osha.gov.
The Government Printing Office website provides a searchable index of the CFRs and can be accessed at http://www.gpo.gov.

Glossary

Balanced scorecard: A balanced set of indicators that link the desired performance results with the operational activities that drive these results. These are often categorized along four dimensions or perspectives of financial, customer, internal business processes and innovations, and learning indicators.

Best practice: An activity or process that represents superior performance.

Beyond compliance: Principle that emphasizes performance over that required by health and safety regulations and laws. It also emphasizes hazard elimination over risk reduction.

Effectiveness confirmation: It is that portion of the audit that evaluates the level of performance of the process against planned results.

Failure mode and effects analysis (FMEA): A systematic and disciplined approach for anticipating failure modes (risks) associated with a process and their likelihood of occurrence, the hazards that cause the risks, and countermeasures to eliminate or control the hazards or reduce the risks.

Job hazard analysis (JHA): Systematic analysis of the hazards associated with tasks and the risk that these hazards present to human operators and equipment.

Nonconformance (NC): An audit finding that represents a violation of a requirement.

Objective evidence: Evidence, typically in the form of records, direct observations, or documentation, of the conformance or nonconformance of a process or activity to a requirement or planned results.

Opportunities for Improvement (OFI): An audit finding that represents, in the judgment of the auditor, an opportunity to improve. Note that an OFI does not represent a violation of a requirement.

Plan-Do-Check-Act (PDCA): A systematic and ongoing process of planning, implementation, measurement of the results, and action to further improve the process. It is also referred to as the Deming cycle.

Registrar: An organization that contracts to perform independent assessments for the purposes of certification of the management system.

Acronyms Used in the Handbook

BSI	British Standards Institute
CAR	Corrective action request
CFR	Code of federal regulations
CFT	Cross-functional team
EMS	Environmental management system
FMEA	Failure mode and effects analysis
HAZOP	Hazard and operability study
HSMS	Health and safety management system
JHA	Job hazard analysis
MSDS	Material safety data sheet
NC	Nonconformance
NCR	Nonconformance report
OFI	Opportunity for improvement
OH&S	Occupational health and safety
OHSAS	Occupational health and safety assessment series
OJT	On-the-job training
OSHA	Occupational Safety and Health Administration
PAR	Preventative action request
PPE	Personal protective equipment
QMS	Quality management system
VPP	Voluntary protection program

About the Author

Joe Kausek is the President of Joe Kausek & Associates, a training and consulting firm in Saline, Michigan. Joe is a retired naval officer, having served twenty years in the Naval Nuclear Propulsion Program. He started his naval career as nuclear operator, where he held positions responsible for plant maintenance, chemistry control, lockout-tagout, confined space entry, airborne monitoring, and the radiation safety and control.

Since his retirement in 1995, Joe has provided quality, environmental, and health and safety management system consulting and training for hundreds of clients in the automotive, defense, aerospace, transportation, health care, and service sectors through Eastern Michigan University's Center for Quality, Baker College, and Joe Kausek & Associates. His areas of expertise include ISO 9001, ISO/TS 16949, ISO 17025, ISO 14001, and OHSAS 18001.

Joe has led over forty companies through the certification process to ISO 9001, QS-9000, ISO/TS 16949, ISO 14001, and OHSAS 18001 and has served as auditor for the Vehicle Certification Agency, a leading international registrar. Joe is also the author of *The Management System Auditor's Handbook* and *Environmental Management Quick and Easy*.

Index